Astrotopia

ΟΤΟΡΙΑ

The Dangerous Religion of the Corporate Space Race

MARY-JANE RUBENSTEIN

The University of Chicago Press

Chicago and London

The University of Chicago Press, Chicago 60637
The University of Chicago Press, Ltd., London
© 2022 by The University of Chicago
Published 2022
Paperback edition 2024
Printed in the United States of America

33 32 31 30 29 28 27 26 25 24   1 2 3 4 5

ISBN-13: 978-0-226-82112-2 (cloth)
ISBN-13: 978-0-226-83338-5 (paper)
ISBN-13: 978-0-226-82317-1 (e-book)
DOI: https://doi.org/10.7208/chicago/9780226823171.001.0001

Library of Congress Cataloging-in-Publication Data

Names: Rubenstein, Mary-Jane, author.
Title: Astrotopia : the dangerous religion of the corporate space
    race / Mary-Jane Rubenstein.
Description: Chicago ; London : The University of Chicago Press,
    2022. | Includes bibliographical references and index.
Identifiers: LCCN 2022011151 | ISBN 9780226821122 (cloth) |
    ISBN 9780226823171 (ebook)
Subjects: LCSH: Religion and astronautics. | Space colonies. |
    Colonies—Religious aspects. | Outer space—Exploration—
    Moral and ethical aspects.
Classification: LCC BL254 .R83 2022 | DDC 215—dc23/
    eng20220611
LC record available at https://lccn.loc.gov/2022011151

♾ This paper meets the requirements of
ANSI/NISO Z39.48-1992 (Permanence of Paper).

Contents

As for those who would take the whole world to tinker
with as they see fit, I observe that they never succeed.

LAO TZU

Preface

It's something out of a bad dream—or a mediocre sci-fi story. Earth is becoming uninhabitable, so a wealthy fraction of humanity hitches a ride off world to live in a shopping mall under the dominion of the corporation that wrecked the planet in the first place. Meanwhile, conditions on Earth approach infernal. Meanwhile, the space colony amplifies every earthly social crisis thanks to unreliable technology, tight living quarters, and the oligarchic control of information, water, and air. We know this story never ends well, but the present moment finds us buying it wholesale from a few charismatic CEOs in agonistic partnership with the big national space agencies. This is the era of NewSpace.

The best-known of the NewSpaceniks is Elon Musk, who infamously wants to "save" humanity from its bondage to Earth. Earth is a disaster, a planetary prison where *Homo sapiens* will eventually meet its nuclear, microbial, or asteroidal end if someone doesn't colonize another planet—and soon. Faced with such imminent obliteration, Musk's aeronautical company SpaceX is aiming to build two starships a week to send a million cosmic homesteaders to Mars.

And Musk is not alone. There's Jeff Bezos, who resigned as CEO of Amazon to pursue his own vision of getting humans off a dying planet. There's the well-funded Mars Society, led by Robert Zubrin, who wants us to hightail it to the Red Planet to "bring a dead world back to life." And there's the US government, making up rules as it goes along and hoping the UN doesn't figure out a way to object. This American astronautic bravado knows no party lines; in fact, the only major Trump-era objectives the Biden administration has retained are (1) the creation of a space force to wage orbital warfare, and (2) the settlement of the Moon and Mars to enact what Donald Trump called America's "manifest destiny in the stars."

As much as ever, the outer space of today is a place of utopian dreams, salvation dramas, savior complexes, apocalyptic imaginings, gods, goddesses, heroes, and villains. As much as ever, there's something weirdly religious about space.

The big argument of this book is that the intensifying "NewSpace race" is as much a mythological project as it is a political, economic, or scientific one. It's mythology, in fact, that holds all these other efforts together, giving them an aura of duty, grandeur, and benevolence. As such, there's not much that's new about NewSpace. Rather, the escalating effort to colonize the cosmos is a renewal of the religious, political, economic, and scientific maelstrom that globalized Earth beginning in the fifteenth century. In other words, the NewSpace race isn't just rehashing mythological themes; it's rehashing Christian themes.

Now when I say Christian, I'm not referring to the Christianity of Martin Luther King, Dorothy Day, Daniel and Philip Berrigan, or any of the other antiracist, antiwar, Earth-loving, creature-nurturing teachers and communities out there—including Pope Francis. Rather, I'm talking about the imperial Christianity, or "Christendom," that teamed up with early capitalism, European expansion, and a particularly racist form of science to colonize the Earth. We can draw an eerily straight line from the "doctrine of discovery" that "gave" Africa to Portugal and the New World to Spain, through the "manifest destiny" that carried white settlers across the American continent, to the NewSpace claim to the whole solar system and, eventually, the galaxy.

In each of these cases, the destruction of the Earth and the exploitation of both human and nonhuman "resources" become sacrificial means toward a sacred end—namely, the wealth and prosperity of a particularly destructive subset of the species. Once the stuff of infinite possibility, outer space has become just another theater of greed and war.

The question, then, is whether there might be a different approach to exploring the universe. Is there a way to learn from other planets, moons, and asteroids without destroying them? Is there a way to see land as important in its own right rather than a mere container for "resources"? Is there a way to visit or even to live on multiple planets without ransacking them? How might we approach outer space without bringing our most destructive tendencies along with us? And might we find ways to heal our ravaged Earth in the process?

What I'm going to suggest is that since the problem is fundamentally "religious," the solution will have to be, too. The challenge for justice-loving space enthusiasts will be to replace the destructive myths guiding our scientific priorities with creative, sustainable, and peaceful ones. Where are these destructive myths? They're everywhere—even grounding some basic assumptions of the modern age that seem to have nothing to do with religion. Such basic assumptions include the idea that intelligence is the most valuable force in the universe, that humans possess more of it than anyone else, that minerals are only valuable as resources, and that land can be owned.

Western science, economics, and politics will tell us that each of these ideas is simply true, even universal. But they are actually the legacy of Western monotheism and Greek philosophy, whose gods and heroes continue to name our spacecraft and "missions." In short, Western science, economics, and politics are most indebted to religion where religion is least obvious.

But there are other traditions, and even other interpretations of Western traditions, that privilege kinship over competition, knowledge over profit, and sustaining over stockpiling. Such other stories are already forming what some scholars call "a new scientific method" that values the knowledge systems of Indigenous peoples, the dignity of nonhuman life-forms, and the integrity of the land— however lifeless it may seem.[1] Rather than extending infinitely the same old mess we've made before, such a "new" approach would aim to create sustainable, just, and joyful communities on Earth— and perhaps even elsewhere.

We Hold This Myth to Be Potential

Another Dimension
Of another kind of Living Life
SUN RA

It's Monday morning and I'm about to tell a screen full of astronomers they should study religion. "What was I thinking," I mutter, climbing over the sprawling limbs of my slumbering four-year-old. After all, these are serious astronomers. They love evidence, hate "blind faith," and fly into an adorable geek-rage when people confuse them with astrologers.

"I'll start with a broad definition," I resolve, smoothing the four-year-old's hair down and checking my phone to make sure his school isn't closed. I start a voice recording so I can remember what to jot down when I finally make it to my desk. "Religion tells us where we've come from, where we're going, and how to live in the meantime. It connects this realm to other realms. It provokes awe at the order of things and horror at the disorder of things." That'll have to do for now; the kid's got my phone.

Origins, endings, worlds, and order. Precise calculations and constant amazement. In this sense, surely the astronomers will agree that science looks a bit like religion and religion looks a bit like science?

This might be harder than I thought.

"Come on, Elijah," I say, shuttling him into his own room, "let's get you dressed before baby Ezra wakes up." We pad together onto his dark blue rug with the white and gray stars. I trip over the miniature astronaut from the wooden space station and throw his rocket ship pillow back onto his bed. Elijah works at taking off his constellation pajamas while I rifle through his drawers to find his space-camel T-shirt. I wrap a moon-and-stars cape around him, hand him his glitter-filled "space scepter," and walk the grinning "space magician" downstairs to pack his galaxy lunch bag into his solar system backpack.

I'm not even exaggerating. All his stuff is space stuff. But here's the thing: Elijah's not all that into outer space. Sure, he likes it, but we're the ones who picked it out for him: his parents, grandparents and godparents, aunts and uncles. We could have just as easily filled his room with dinosaurs, unicorns, Batman, or Daniel Tiger, all of which are no less exciting to him than Mars or Saturn—and no less real. As far as Elijah's concerned, there's no qualitative difference between astronauts, Octonauts, aliens, Mercury, Gotham, Mrs. Claus, Ganesh, George Michael, and Elena of Avalor. All of them are both not quite real and superreal; they occupy most of his thoughts and guide his actions even though he's never actually met the Hulk or been to Jupiter. As such, all these characters and realms are, for him, what a religion nerd like me would call *mythic*.

Why space stuff, then, when it could have been sharks or trains? What is it about space that makes me—makes *us*—want our kids to get excited about it? What are the values of "space" that we're hoping our little humans will pick up?

"Space Is the Place," chants the Afrofuturist jazz musician Sun Ra, along with his cosmic Arkestra. What Ra means is that space is the place of new life, new consciousness, and a different kind of harmony. The place for the displaced, the placeless; those whose place is not here. For Ra, space is the place for those who've had enough of a planet run by oppressive, profit-driven, slave-driving warmongering—and those who are ready for a new way to be. And so, he intones, declaring our interdependence,

We hold this myth to be potential
Not self-evident but equational
Another Dimension
Of another kind of Living Life[1]

Maybe that's it: even for grown-ups, the myth of space remains a myth. Long after Santa and Elmo are gone—even if Ganesh and Jesus are gone—space still amazes us, opening what Ra calls, simply, *potential*. When we teach our kids to love planets and stars and interstellar blackness, we teach them to love infinity, expansiveness, and not-quite-knowing. To linger with the puzzles that elude them; to watch out for something genuinely new and maybe even become it.

The place of our most poetic imaginations and our most obsessive calculations, space is the place where art, science, literature, technology, and religion all attract and repel one another in a vortical frenzy, promising this or that path toward enlightenment, that or this more perfect existence. Space consistently hits us with shock and calls forth our awe, telling us, for example, that our glorious, all-giving Sun is just one of *hundreds of billions* of stars in the Milky Way, which is one of hundreds of billions of galaxies in our observable universe, which may or may not be one of an infinite number of universes. Innumerable suns warming scadzillions of planets, with oceans and dust storms and cloud microbes and who knows what else, all in constant motion through infinite space and time, and here you are, making a cheese sandwich, nowhere in particular.

*   *   *

But while some of us are lost in wonder, obsessed with understanding, or committed to cosmic justice, others are using the ineffable pull of outer space toward nationalist, military, and increasingly commercial ends. In this NewSpace age, public and private interests are both cooperating and competing to build permanent outposts on the Moon, mine water and metals from planets and asteroids, and eventually colonize Mars—all under the seemingly noble auspices of "fulfilling our destiny," building a "clean, green future," and even "saving humanity." How will the astrosaviors *accomplish* such deliverance? By converting the cosmos itself into capital and conquering space, the final frontier.

Salvation through imperialism. As far as outer space goes, the strategy is as old as the Apollo era, when the US planted an American flag on the Moon "for all humanity"; claimed American military supremacy in the name of world peace; and circulated those orbital photos of a beautiful, unified Earth to announce the birth of environmentalism on the one hand and of global finance on the other. Farther back in history, such salvific conquest can be heard in the "Manifest Destiny" that called white Americans across the Western frontier while displacing Native Americans and destroying their land. And farther still, in the "Doctrine of Discovery" that lent divine authority to the European seizure of the Americas, the murder of First Nations, and the enslavement of African people. In all these frontier-driven journeys—across the seas, over the North American continent, and now out into space—commercial interest and national glory have been secured by means of massive human and ecosystemic suffering. But accounts of this suffering are consistently drowned out by heart-lifting appeals to prosperity, destiny, salvation, and freedom.

So this is why I'm worried about NewSpace. In their promises to get a few humans off this doomed planet, billionaire utopians are selling the same old story of domination hidden under lofty religious language.

The central character in the astrotopian drama is Elon Musk, CEO of SpaceX and the perennial rival of Jeff Bezos for the title of richest man in the world. Having made his fortune through PayPal and Tesla, Musk now spends most of his prodigious energy on his aerospace company, whose stated mission is to make humanity a "multiplanetary species," beginning with a colony on Mars. At stake, says Musk, is the future itself: whether it's an asteroid, nuclear war, mutinous robots, or a really nasty virus, something is bound to wipe out life on Earth soon, so the survival of the human species depends on our getting the hell off this planet.

Hence his self-righteous response to Senator Bernie Sanders, who accused both Musk and his rival billionaire Jeff Bezos of perpetuating a "level of greed and inequality" that was at once "immoral" and "unsustainable." After Musk-apologist site CleanTechnica annotated Sanders's tweet with a curved arrow pointing toward the word "greed" and five angry, orange question marks, Musk shot back, "I am accumulating resources to help make life multiplanetary and extend the light of consciousness to the stars."[2] Money has nothing to do with it; Musk is on a mission.

Musk's followers are fervent, numerous, and defensive. As one audience member shouted after he announced his cosmic intentions, "[Musk] inspires the shit out of us!"[3] And although few of these Musketeers would describe themselves as religious, they have bought into a classic myth of disaster and salvation delivered by a self-appointed savior. "This world is coming to an end," the savior cries, "but trust in me and I'll bring you to a new world, where you'll finally be free." Free from death, free from Earth, free from gravity—at least most of it—and even free from international regulation. (Read the fine print of Musk's Starlink contracts, and you'll see him declare Mars "a free planet," over which "no Earth-based government has authority or sovereignty."[4])

Living seven months away from their birth planet, Musk's first Martian pilgrims will be on their own, building a brand-new, "self-sustaining" society in an underground cave with the power of hard work, robots, and a fleet of indentured servants. Eventually, these high-tech homesteaders will be able to live on the planetary surface, having "terraformed" Mars to be as Earthlike as possible. How do you take a freezing, radioactive, blood-boiling planet and make it like our life-loving Earth? Well, says Musk, you "warm it up," which he proposes to do by dropping some atomic bombs on it—hence the "Nuke Mars" T-shirts for sale on the SpaceX retail website.

Musk's cosmic messianism would be easy to dismiss if SpaceX weren't actually sending thousands of satellites into space, selling civilian rides on low-orbital rockets, and carrying cargo and astronauts for NASA. In fact, thanks to recent legislation, the American space agency is increasingly dependent on the better-funded, less-risk-averse "astropreneurs" to equip and execute its missions. Billionaire spaceniks like Musk, Bezos, and a host of space-mining corporations are now competing viciously for enormous government contracts.

In the meantime, China's got a rover on Mars, a satellite on the far side of the Moon, and an independent space station in the works. Richard Branson is selling $200,000 tourist tickets on his notoriously unreliable space plane. Space-mining firms have collected billions before they've even gotten their hands on an asteroid. A failed Israeli mission has dumped dehydrated tardigrades on the lunar surface. An increasing number of African nations are ramping up their own space programs. The US has created a new branch of the military to wage extraterrestrial war. And some astronautic start-up says it's going to build a Ferris wheel–shaped orbital space hotel.

Meanwhile, Earth is surrounded by the metal shards of every dead satellite, botched mission, and detached bolt in history, encircling the planet in a dense "corona of trash" that no one seems to know how to get rid of.[5] Space is an absolute mess.

It's frankly enough to make the parent in me want to yell "cut it out"—as if anyone would listen. It's enough to make the ecologist in me terrified that, having trashed one world, we're storming off to ransack others. And it's enough to make the humanist in me want to beg Musk and Bezos to spend their obscene fortunes on things like water access, biodiversity, education, reparations, and reforestation so we might avoid the doomsday scenarios that fuel their escape fantasies in the first place. It's enough, in short, to make me want to give up on space.

But then I think back to the astronomers I'm about to perplex with my lecture on religion. To their wonder at the sublimity of the universe they so monkishly study. To their dazzling decoding of the beginnings and nature of things. To the way I cringe empathetically when I hear them resorting to utilitarian defenses of their research programs, as if knowledge is only important if it cures Covid or speeds up cell phones. I think of Sun Ra, Janelle Monáe, and those feminist, Indigenous, and Afrofuturist sci-fi authors who find in "space" the possibility of a genuinely free future for dehumanized people. And I think of my little space wizard at home in his planet tent and realize that, honestly, I'd *love* it if he and his brother became astronomers. Or cosmic poets. Or futurist jazz pianists singing better worlds into being. But it's going to be a hard sell if the planet's being strangled by a garbage halo.

* * *

In recent years, a small but tireless group of scientists has begun to insist that we have to do space differently. Justice-loving astronomers have joined forces with performance artists, musicians, philosophers, activists, and anthropologists to call for a "decolonial" approach to observation and exploration. The academic language can be confusing, but "decolonizing space" would mean diversifying the astronautic industry along the lines of race, gender, and class—not just for the sake of representation but in order to approach outer space from as many perspectives as possible. Decolonizing space would mean centering Black and Indigenous voices in all plans concerning extraterrestrial labor and territory, which must not be romanticized as "hard work" and the "empty frontier." It would also mean refraining from polluting other planets (and the interplanetary spaceways), refraining from extracting "resources," refusing to commodify land, and subjecting private enterprises like SpaceX and Blue Origin to strict national and international regulation.

In this way, these activists suggest, space might escape its romantic but sinister designation as "the final frontier." After all, the story of the earthly frontier—especially in the Americas—is violent, genocidal, and ultimately ecocidal. So rather than resurrecting this old exploitative story under the guise of heroism, they suggest that space could be a place of genuinely new beginnings and reimagined relationships—"to other forms of life, to land, and ultimately to each other."⁶ I want so badly to believe them. In the military-commercial throes of the quest to conquer space, I want to be able to say, alongside the decolonial astronomers, that another space is possible. But to get there, we're going to have to unearth the old, destructive myths behind the escalating NewSpace race and let other myths guide us.

Our journey begins in the next chapter with the current state of affairs in and around outer space. Over our heads, behind our backs, and under our feet, private enterprise is both aiding and competing with national interest to establish a permanent human presence and a cutthroat economy beyond our Earth. In the US especially, these endeavors are rhetorically justified with the language of destiny, freedom, salvation, and even divine will. For this reason, our exploration will progress by turning backward, to those biblical stories and dogmatic teachings that justified European colonialism, US expansionism, and the Cold War Space Race. We will encounter the political, ethical, and environmental crises resulting from this religiously justified imperialism, which are now culminating in the militarizing of space, the race to exploit extraterrestrial "resources," and the dizzying proliferation of techno-garbage orbiting our planet.

As we will see, these crises are escalating faster than science and legislation can run. It is therefore crucial to listen to the voices of people with other ideas. By turning to Indigenous philosophies, more-than-human worldviews, religious ecologies, and Afrofuturist visions of extraterrestrial justice, we can begin to imagine communities of genuine equality, peace, and freedom based on respect and even reverence for both living and nonliving beings.

What I'm going to suggest is that if we want to get right with space, we're going to have to get right with religion. To expose the values of contemporary techno-science as the product of bad mythologies and seek out better ones. To find—and even make—stories that put caretaking over profit and harmony over ownership. Stories that tell us not how the universe might belong to us but how we might belong to the universe.

Our Infinite Future in Infinite Space

Our destiny is to become the gods that we
once feared and worshipped.
MICHIO KAKU

The Hare and the Tortoise

I first realized something was up when Elon Musk launched a car into orbit. It was January of 2018, and SpaceX was looking to test its Falcon Heavy rocket, woo the US military, and make sure everyone was watching. So rather than display the rocket's carrying capacity with conventional slabs of concrete or steel, Musk decided to strap a blazing red Tesla Roadster to its back. A perfectly good, even exquisite, car. One hundred thousand dollars' worth of chrome, leather, steel, glass, state-of-the-art navigation software, green technology, and human labor hurled into useless orbit—not around Earth but around the *Sun*. It was an act of immense bravado, extraordinary waste, and literally cosmic presumptuousness: now along with eight planets and some dwarves, moons, and asteroids, there is a tricked-out convertible circling our solar orb, driven till the end of days by a mannequin in a space suit called Starman.

Musk named his doomed astrobot after the alien messiah of David Bowie's 1972 *Ziggy Stardust* album. Bowie's "Life on Mars" accompanied the Falcon Heavy launch that flung the Roadster toward the stars, and his "Space Oddity" still loops endlessly on its

Figure 1.1   Starman in his orbital Tesla.

SpaceX

JVC speakers. Starman's glove compartment is stuffed with multi-media versions of Douglas Adams's *Hitchhiker's Guide to the Galaxy* and Isaac Asimov's *Foundation* trilogy. And the rocket itself is named after the Millennium Falcon of Star Wars. Musk, you might say, is a geek's geek, his aesthetic composed by nostalgia for the future of his teenage past: rockets, space suits, Martian colonies, glam rock, and the free market promise of infinite possibility.

Musk is also an inveterate showman. Back in 2003, he was having a hard time getting NASA to take SpaceX and its newly fabricated Falcon 1 seriously. So he drove the seven-story rocket on an enormous flatbed truck from El Segundo, California, to Washington, DC, and parked it on the street outside the headquarters of the Federal Aviation Administration.[1] In the two decades since, Musk has continued to manufacture all manner of eye-catching spectacles: Twitter-stormed launches, dramatic explosions, tickets sold to billionaires for trips on unbuilt ships, promises to rename the city of Boca Chica, Texas, "Starbase," and a manifesto about his intention to save "humanity" by getting the hell off the planet Earth.[2]

Meanwhile, on the other side of Texas, Jeff Bezos has been making a lot less noise. In the early 2000s, while Musk was filing antitrust lawsuits against NASA, the air force, Boeing, and Lockheed

Martin, Bezos was quietly buying up ranches. Under the auspices of improvised corporations named after legendary frontiersmen (John Cabot, James Cook, William Clark), Bezos cobbled together over three hundred thousand acres of land in West Texas so he could test his rockets without anyone noticing.[3] Musk bought land, too, of course, but he makes so much noise that the rangers at Mother Neff State Park near Waco now warn their visitors that if something sounds like the end of the world, it's probably not. (At least not yet.)

The two men are different sorts of magicians. Musk pulls rabbits out of hats while Bezos makes the coin disappear behind your ear. While Musk is shouting, "Look, Mom! Wait, that's not it," Bezos hides in his room to perfect the trick. Both men are building reusable, affordable, state-of-the-art rockets, but Musk raced to the launches while Bezos worked on the landings. Elon's had us looking up at the skies while Jeff's kept us staring at our own laps, 1-Clicking the lint rollers, cake pans, and dog sweaters that finance his more cosmic endeavors. "Every time you buy shoes [on Amazon], you're helping Blue Origin," Bezos explained when he finally revealed what he was up to in West Texas; "I appreciate it very much."[4]

Bezos himself articulated the methodological difference between Blue Origin and SpaceX in a 2004 letter to his then-tiny aerospace staff: "Be the tortoise," he told them, "and not the hare."[5] His motto for the company is *Gradatim ferociter*, or "step-by-step, ferociously"—a grittier, Latinate rendition of "slow and steady wins the race." The phrase is inscribed on banners beneath the company's crest (it's got a crest), which features two turtles standing on top of a globe, reaching from North America to a gilded solar system. Crowning the image is a cruciform sun; anchoring it is a winged hourglass with all the sands run out; and the whole thing looks as if a fifteenth-century cosmography walked into a Harry Potter fanzine.

More nerd than geek, Bezos reads everything in print, considers even the most outlandish alternatives before making up his mind, and demands that ideas be pitched in full-paragraph form. As we all know, he's a books guy; in addition to Aesop, his references include

Tolkien, Asimov, Verne, Ian M. Banks, Neal Stephenson, and William Gibson. Now and then, he even mentions *A Wrinkle in Time*, one of the only works by a woman author to make the astropreneurial canon. He tells aspiring tycoons to read the personal accounts of every famous CEO in recent history. But when it comes to space, Bezos's biggest influence is *Star Trek*.

While Musk is off actualizing George Lucas with his exploding Falcons and epic soundtracks, Bezos is cultivating the more genteel gestalt of the starship *Enterprise*. As the *Atlantic*'s Franklin Foer reports, Bezos initially wanted to call Amazon "MakeItSo.com" as an homage to Captain Jean-Luc Picard, whom he now uncannily resembles.[6] He named his dog Kamala in honor of the empathic metamorph from Krios Prime, who is said to be Picard's "perfect" yet unattainable "mate."[7] One of the secret corporations buying all that desert in Texas was called Zefram LLC, after the human scientist who enabled contact with the Vulcans by discovering warp drive (even turtles like to break the speed of light from time to time). And in the fall of 2021, Bezos boldly took William Shatner where no ninety-year-old actor had gone before.

The question you might be asking is, *Why?* What are these billionaires up to in space?

A Tale of Two Utopias

LIFE ON MARS

It's probably old news to you by now: Elon Musk and Jeff Bezos want us off the planet. Not all of us, of course, but according to both of these absurdly wealthy utopians, the future of the species will depend on those humans who've got the foresight, fortitude, and finances to head to outer space. Just as it was for Captain Kirk and the Apollo crews, space has become for these latter-day pioneers "the final frontier": a place of new worlds, untold fortunes, and immense danger. It is no coincidence that the "ultraelite" Explorers Club in Manhattan chose to honor both Musk and Bezos at the same dinner in 2014. Over a characteristic danger-menu at the Waldorf Astoria featuring such delicacies as "goat and goat penis" and alligators on spits, the club credited both men with "revolutionizing both space exploration and sustainable transportation."[8] As Bezos rose to accept his award, he reportedly said, "I'm still making sure there's no cockroach in my teeth."[9] So there's the peculiar masculinity of the new corporate space race: classical sci-fi fan meets gamer meets Rockefeller meets Indiana Jones.

Musk and Bezos are notorious rivals, competing for contracts, exchanging online jabs, and continually trading places as the wealthiest man on Earth. Both of them have testified repeatedly that their efforts in space are of utmost personal, professional, and existential importance, and that their obscene fortunes are justified as means to a humanitarian end—which is to say, the salvation of life as we know it. Despite these significant similarities, however, the two men differ not only in temperament and approach—as we have seen, Jeff Bezos plays the self-proclaimed tortoise to Elon Musk's hare—but also in values and vision. When it comes down to it, the two billionaires want different things in different places for vastly different reasons.

Infamously, Musk wants to go to Mars. In fact, as he explained in a 2016 manifesto, it's been his goal all along: "making humans a multiplanetary species" by setting up a "self-sustaining city" on the Red Planet.[10] Having learned the argument from the aerospace engineer and longtime Mars advocate Robert Zubrin, Musk explains to anyone who will listen that Earth is a ticking time bomb. Sooner or later, something will destroy humanity, whether it be an asteroid, nuclear war, or AI robots gone rogue (he's particularly worried about whatever Larry Page is up to over at Google). We're therefore going to have to find somewhere else to live, and given the literally infernal conditions of Venus—the average temperature is around 850°F (455°C) and the clouds "rain" sulfuric acid—Mars is our best chance. Of course, five billion years from now, the Sun will explode into a red giant and engulf Mars along with Earth in a fiery apocalypse. So if we want humanity to endure forever, we'll eventually have to make it to another solar system—the more systems the better, from an evolutionary perspective. But we'll never be able to live anywhere else unless we start close to home—and soon, before a giant asteroid or Alexa 5.0 wipes out the whole species.

At times, Musk seems to realize how much he sounds like that guy on the street with a cardboard sign that says "THE END IS NEAR." Both disavowing and adopting the role of lunatic (sorry, Martian) prophet, he writes, "I do not have an immediate doomsday prophecy, but eventually . . . there will be some doomsday event." With apocalypse on the horizon, our first option is to let the disaster extinguish us as it did the dinosaurs—an option Musk finds so intolerable he never even entertains it. "The alternative," he says, "is to become a space-bearing civilization and a multi-planetary species, which I hope you would agree is the right way to go."[11] So Mars it is.

Much like any other doomsday prophet, Musk keeps revising his timeline. Having initially promised to send crewed missions to Mars in 2020 and then 2025, Musk now hopes to send the first few humans to the Red Planet just before or after 2030, with the goal of getting a million people to Mars by 2050. The challenge will be making the enterprise affordable . . . ish. Using conventional technology, the price for a round-trip ticket to Mars would currently be $10 billion a person. Once his rockets attain full reusability and efficiency, however, Musk predicts he'll be able to lower the cost to $200,000, "the median cost of a house in the United States."[12] At that price, he reasons, "almost anyone" could go to Mars. All they'd need to do is save up a bit, sell their house if they've got one, and pack a (very small) bag.

Fashioning himself after the American tycoons of the nineteenth century, Musk promises to build an interplanetary transportation system akin to the transcontinental railroad. This cargo route will bring earthly supplies to the nascent Martian colony every twenty-six months, when the two planets come closest to one another. As it becomes self-sufficient, the colony will rely less and less on these deliveries, somehow gaining the capacity to grow its own food, manufacture its own fuel, and mine sufficient resources to create and sustain infrastructure. Eventually, there will be no need for ships to come at all, except to transport passengers and perhaps to engage in trade.

When it comes to advertising his new colony, Musk alternates between appealing to aspirational homesteaders and revving up postprom kids. On the one hand, he admits, Mars is going to be seriously hard work. Under current conditions, it is impossible to breathe or even just *be* on the planet without a space suit. Since Mars has so little atmosphere, it would turn all the water in a human body to steam and kill it instantaneously. Even with a space suit, there is so much radiation on Mars that it will likely cause the colonists severe health problems. So as Musk concedes from time to time, Mars will be like the Oregon Trail on a really bad day: "There's a good chance you'll die; it's going to be tough going."[13]

On the other hand—and this is the part Musk tends to dwell on— the Martian Trail is going to be pretty cool. The trip itself will be like an astronautic Club Med: a hundred people on one four-hundred-foot "Big Fucking Rocket" on a seven-month trip that will never "feel cramped or boring." There will be zero-gravity games (Musk is really into bouncing around), plus "movies, lecture halls, cabins, and a restaurant. It will be really fun to go," Musk enthuses; "you are going to have a great time!"[14] (Nowhere in these descriptions does Musk explain who will be staffing the restaurants, cleaning the cabins, or wiping the space puke off the gleaming walls of the BFR. Once in an interview, he suggested that people who couldn't afford the $200,000 trip to Mars might work to pay it off, so maybe that's where he'd find his custodial class: in a dedicated fleet of indentured servants.[15])

As for the planet itself, Musk promises, "it would be quite fun to be on Mars, because you would have gravity that is about 37% of that of Earth, so you would be able to lift heavy things and bound around."[16] Sure, the air is primarily carbon dioxide, but the same stuff that's waste material to humans will make it easy to grow plants "just by compressing the atmosphere." Faced with the problem of "the radiation thing," Musk says inexplicably that it's "not too big of a deal,"[17] and although he understands that Mars is "a little cold"—the average temperature is –80°F (–62°C), which is to say over 100 degrees below freezing—he assures his future colonists that "we can warm it up."[18]

How exactly do you "warm up" a frozen planet? Musk's ideological predecessor Zubrin proposes "terraforming" Mars by "greenhousing" it; that is, imitating the process currently roasting Earth by releasing halocarbons (chlorofluorocarbons; CFCs), genetically engineered gassy bacteria, or even more carbon dioxide into the atmosphere of Mars.[19] The popular physicist Michio Kaku favors the idea of harvesting methane from Saturn's moon Titan and importing it to the Martian skies.[20] But all this sounds far too complicated to Musk, who suggests we can just "nuke Mars," instead. Hit the airspace above the ice caps with some hydrogen bombs, and you'll jump-start the warming process, liberate tons of water, and move the colony that much closer to autonomy.

To be sure, most scientists think this is an absolutely ridiculous plan. Astrobiologist Lucianne Walkowicz warns that the atmosphere of a thermonuclearly warmed Mars would still be too dry to rain, and what's worse, it would have become toxic to any imaginable form of life. Even if it were desirable, it's not clear that it's *possible* to terraform Mars by nuking it; the director of the Russian space agency Roscosmos has estimated it would take more than ten thousand missiles to carry out Musk's nuclear option.[21] Elon's Twitter response? "No problem."[22]

Thanks to some bold engineering and tireless iteration, Musk promises that Mars will eventually resemble Earth, with rivers, lakes, trees, shopping malls, and video games. The only major difference is that we'll be able to jump a lot higher and throw a lot farther.

If you're furrowing your brows at this fantasy of planetary hacking, you're not alone. As Walkowicz reminds us, we don't have a great track record of controlling biotic processes on the planet we've already got. How can we hope to make a habitat out of Mars when we can't even preserve the habitability of Earth?[23] It would seem that regulating the biosystems of an already-oxygenated, temperate, blue-green orb would be a far easier task than bringing a planetary dust storm to life. I mean, we can't even figure out how to prevent a few devastating degrees of climate change on Earth.

Actually, it's not that we don't know how to do it. It's just that we don't want to.

When asked why he wants to "save" humanity by sending us to Mars rather than by addressing injustice, poverty, and climate change on Earth, Musk will often laugh and say, "Fuck Earth."[24] Earth is done; Earth is history; Earth is *so* last eon. Considering the coral reefs, wetlands, and clean skies that SpaceX has polluted and destroyed—and considering Musk's own advancement of the artificial intelligence he worries might wipe us out—one could even accuse him of worsening the disaster to intensify the need for salvation. Of making the planet genuinely uninhabitable so that we will, indeed, need to leave it. In this light, Musk resembles Friedrich Nietzsche's caricature of a priest, who heals his flock by making them sick in the first place.[25] Just as a priest can only offer you salvation from sin by convincing you you're sinful, Musk is urging us to get off this doomed planet by escalating our planetary doom.

Although I hesitate to psychoanalyze a man I've never met, it might be instructive to know that Musk grew up as a rich, white child under South African apartheid.[26] As soon as he was old enough, he left his "impoverished and racially divided country" and moved to Canada and then America, that perennially mythic land of new beginnings.[27] Something similar seems to be going on with his aspiration to leave Earth behind and make a fresh start on Mars. For all its uncertainty and even hostility, Mars is *new*. On the New America of Mars, we have the chance to escape the problems of the past and start again. On Mars, we might finally build a utopia.

The word *utopia* comes from the Greek word *topos*, or "place." The *u* is privative, meaning that it negates the word it precedes. Etymologically, then, *utopia* means "no place." And it's just this imprecision, this lack of location, this perpetual fuzziness that allows utopianism to flourish. If it's never quite anywhere, or never quite realized, then a utopia can be whatever you'd like it to be.

Classic utopians like Plato, Thomas Moore (who coined the term), and Marx and Engels gave us very clear ideas of what their ideal societies would look like: classes either concretized or demolished, money either distributed or abolished, and so forth. Musk, by contrast, offers what you might call a utopianism without the utopia. You won't find any social or political blueprints in his motivational talks or business plans. What you'll find instead are abstract promises of "freedom"—from Earth, from international regulation, from gravity, and even from death, at least at the level of the species. He hasn't hammered out the details, because the details would destroy the perfection. But it's going to be awesome on Mars.

SITTING IN A TIN CAN

Jeff Bezos isn't so sure. In fact, he thinks Mars will be perfectly awful. We've sent probes to every planet in the solar system, he reasons, "and believe me, Earth is the best one. There are waterfalls and beaches and palm trees and fantastic cities and restaurants. . . . And you're not going to get that anywhere but Earth for a really, really long time."[28] Like Carl Sagan before him, Bezos believes our journeys to space reveal the rare beauty of our "pale blue dot, the only home we've ever known."[29]

Elon Musk knows his Sagan, too, but rather than revering the cosmic master, he seeks to one-up him. In a recent interview, Musk reads a soaring, introspective Sagan passage only to interrupt and revise it: "Our planet is a lonely speck in the great enveloping cosmic dark," Musk intones, his affect flat; "There is nowhere else, at least in the near future to which our species could migrate."

"This is not true," Musk declares, looking up from the book and laughing like it's obvious. "This is false. Mars."

"And I think Sagan would agree with that," the interviewer croons back, "he couldn't even *imagine* it at that time."[30]

Jeff Bezos would not agree that Sagan would agree. "To my friends who want to move to Mars one day," says Bezos, "I say, 'Why don't you go live in Antarctica first for three years, and then see what you think?' Because Antarctica is a garden paradise compared to Mars."[31] Sure, our knowledge of the solar system may have grown since Sagan's day. In the intervening years, we've learned that there's water on the Moon, there may have been life on Mars, and there might be microbes in the poison clouds of Venus. But for Bezos, the lesson remains: Earth is unique and must be preserved at all costs. So if Musk is happy to "fuck Earth," Bezos is set on saving it; if Musk named his SpaceX after the place he'd like to go, Bezos named Blue Origin after the place he'll always be from: this "gem" of a planet called Earth.[32]

How, then, will Bezos restore and preserve the blueness of our origin? The beauty of our Earth? By getting us the hell off the planet. Like Musk, Bezos begins his own story with impending disaster, but his opening character is Earth's ecosystem rather than the human species. While Musk worries about the earthly disasters that might obliterate humanity, Bezos worries about the increasing strain humans are putting on Earth itself. The problem, for Bezos, is energy: we're using too much of it. Given an expanding and "modernizing" human population, global-industrial humanity will reach some absolute limits within the next century. There is simply not enough fuel—whether from the ground, the wind, or even the Sun as it's accessible to earthlings—to power a whole planet's worth of first-rate hospitals, bleeding-edge electronics, megachurches, superstores, slaughterhouses, and industrial farms. We need more energy, so we've got to go to space.

More philosophically minded than Musk, Bezos pauses to consider a few objections. Efficiency won't save us, because no matter how many solar panels or LED bulbs we install, our Earth and its resources are stubbornly finite. The only real alternative would be to stop using so much energy, but that would require "rationing" and perhaps even "population control," both of which Bezos finds intolerable. "These things are totally at odds with a free society," he says. But the biggest problem with the prospect of sustainable life on Earth, says Bezos, is that "it's going to be dull. I want my great-great-grandchildren to be using more energy per capita than I do.

And the only way they can be using more energy per capita than me is if we expand out into the solar system."[33]

So that Marxist adage of the early nineties is true: it's easier to imagine the end of the world than the end of capitalism.[34] Rather than proposing an alternative to the extraction of "resources," the relentless pursuit of profit, and the wasteful cruelty of factory farming; rather than using his prodigious intellect to solve the problem of food distribution or his prodigious fortune to seed a universal basic income (or to pay a few dollars in federal taxes), Bezos is spending his money and time exporting the whole damned system into space. The alternative would be "stasis," or even reversal. And Bezos wants to keep moving "forward," so he's going to have to go up and away.

Here's the way it will work. Rather than schlepping all the way to Mars, Bezos proposes that we construct a series of bases on the Moon. We install solar panels on every base, gaining access to far more solar energy than we can ever harness on Earth. We mine the Moon for water, whose elements can be split and recombined into rocket fuel. Using far less energy than we'd need at Cape Canaveral (the Moon has a lower escape velocity than Earth), we can power minimissions to mine asteroids for heavy and rare-earth metals, at which point we begin to construct miles-long, free-floating habitations between Earth and the Moon.

Yes, you've read that right: giant space pods. The idea comes from Gerard O'Neill, the Princeton physicist who began proposing in the mid-1970s that all heavy industry and much of the human population be moved into space. Mines and factories would occupy asteroids and the Moon, while residence, recreation, and commerce would take place in giant cylindrical tubes, rotating to simulate gravity and positioned at "Lagrange points" to maintain a steady orbit. Having attended O'Neill's lectures in college, Bezos remains a devotee. "This is Maui on its best day, all year long," he promises of his space pods. "No rain, no storms, no earthquakes."[35] In our climate-controlled Edens, we'd have everything we love on Earth, like air, trees, birds, and beaches, but nothing we hate—

Figure 1.2 Toroidal colony, exterior view.

NASA Ames Research Center

O'Neill infamously promised we'd finally be free of mosquitos. And in the meantime, Mother Earth will get a long-overdue nap.

With all heavy industry and a good deal of humanity relocated off-planet, Earth can be zoned for light industry, some residence, and recreation. In short, Earth becomes a planetary park—a great vacation spot, a lovely place to go to college.

Meanwhile, out in space, humans get to play as many video

Figure 1.3 Toroidal colony, interior view.

NASA Ames Research Center

games, have as many kids, and eat as many alligators and goat genitals as they'd like—now that they've got limitless energy. According to Bezos's calculations, an O'Neill-hacked solar system could in principle support one trillion human beings. "That's a thousand Mozarts. A thousand Einsteins," he reasons. "What a cool civilization that would be."[36]

It seems a cheap but necessary shot to point out that, by this logic, we would also gain a thousand Hitlers and Stalins. But Bezos is leaving it to the STEM kids he strategically places at the front rows of his lectures to work out the details. How are we going to build O'Neill colonies? Out of what materials? Under what sort of political systems? Bezos has no idea. He's here to build the infrastructure so that the big thinkers of the future can hammer out the details.

Take Mark Zuckerberg, for example (Zuckerberg is always

Bezos's example). He and his roommates were able to build Facebook, which has revolutionized everything from advertising to social relations, because the infrastructure was already in place. The internet was up and running, power companies were lighting up their dorm room, the university system was educating and socializing them, a small army of workers was feeding them and cleaning their spaces, and roads and bridges delivered constant supplies to keep them warm and alive. (I'm extrapolating. When Bezos refers to the "infrastructure" that supported the creation of Facebook and even Amazon, he stops at the internet, the banking system, and the US postal system, forgetting to mention the transportational, custodial, and nutritional infrastructure supporting this infrastructure. But it seems important to point out that both Bezos and Zuckerberg needed all of these social and industrial services to build their "self-made" fortunes.)

In short, what Bezos is trying to do in space is to prepare the way of the astro-Zuckerbergs. To give them the infrastructure they'll need to do things we can't even imagine. Toward that end, his team has designed a suborbital rocket named after Alan Shepard and an orbital rocket named after John Glenn. New Shepard is already taking tourists on an eleven-minute trip into weightlessness, and New Glenn will "build the road to space" by delivering massive payloads into orbit. A lunar lander nostalgically named Blue Moon will deliver enough people and supplies "to enable a sustained human presence on the Moon."[37] In short, Bezos will establish the extraterrestrial roads and bridges so that future entrepreneurs can figure out what to do with them. Jeff will pave the way for future Jeffs and Marks—and even future Elons, once they've had enough of those radioactive dust storms on Mars.

So these are the two utopias: "fuck Earth and occupy Mars" versus "save Earth by drilling the universe."

And the public is getting excited. As offhandedly "anticorporate" as your average middle-class American may profess to be, we quite like our fast cars and same-day deliveries, especially if they make us think we're doing something virtuous. As one college newspaper puts it, Elon Musk and Tesla are "saving the planet by being awesome."[38] And as Foer reports in the *Atlantic*, Americans express "greater confidence" in Amazon.com than in "virtually any other American institution," including the military.[39] Order a three-pack of airtight canisters and you get, the next day, a three-pack of airtight canisters. Figure out how to open the doors of a Tesla, and that thing will get you two hundred miles away on one charge while accelerating like a dream, stopping on a dime, recommending local restaurants, and entertaining your passengers with fart jokes and video games. Bezos and Musk have built companies that *work*. Why not trust their visions of our future in space?

Of course, both these visions are a long way off. So far, no one's been to Mars, no one's mined an asteroid, and it's been half a century since anyone walked on the Moon. But in the meantime, the NewSpaceniks are already making a total mess. Musk has filled his allotted altitude in low Earth orbit with so many Starlink satellites that he's edging into the territory allocated to Amazon's aspirational Kuiper satellites, insisting that since Amazon isn't yet using the space, Starlink should be allowed to take it.[40] Astronomers and space ecologists keep warning that between dead satellites, live satellites, paint chips, shrapnel, and the International Space Station, there's just *too much stuff* up there. At speeds of eighteen thousand miles per hour, the collision of anything with anything else is disastrous, and despite our ability to produce this deadly litter, we have no reliable way to clean it up. (The most promising idea so far is that we might be able to snag some passing garbage with a harpoon. A *harpoon*.)

It's total chaos, and yet Bezos, Musk, and a growing cadre of smaller-time astropreneurs continue unfazed, promising thousands more satellites, suborbital tourism, orbital tourism, private space stations, space hotels, and kazillion-dollar asteroids, all as means to our beautiful future in space. The road to utopia is paved with towering egos and careening space junk.

At this point one might begin to wonder whether there are any grown-ups in the room. Is anyone in charge?

Private Space

The short answer is, not really. "Space" is the most recent arena of massive deregulation and privatization under the reigning economic strategy known as neoliberalism. It's a long, dramatic story, filled with lawsuits, countersuits, and frustrated rich guys, but for our purposes there have been two critical developments during the first quarter of the twenty-first century.

The first was President Barack Obama's decision to cancel the Constellation mission that was supposed to succeed the moribund space shuttle. Advisers both within and outside NASA had grown tired of the reportedly slow pace and massive expense of the federal space program. As Kaku charges, NASA had been "spinning its wheels for decades, boldly going where everyone has gone before."[41] So in an address delivered on "Tax Day" of 2010 at the Kennedy Space Center, Obama announced his intention to stimulate the space industry by handing it over to the private sector. Not all of it, of course, but enough to "accelerate the pace of innovations" by shifting the burden off the taxpayer and laying off thousands of federal employees.

Although Obama expressed regret that these workers would lose their jobs, he was confident they'd find work elsewhere—namely in the burgeoning private space industry. With more companies competing for government contracts, he reasoned, privatizing space would drive down costs, "revitalize NASA and its mission," and save the US from having to rely on Russia to get its astronauts into orbit. And although Obama announced this decision more than a year into his first term as president, the shift toward deregulation and private industry was already well underway. Even before Obama took office, SpaceX had already been appointed a major beneficiary of the impending transformation, winning $1.6 billion

Figure 1.4 President Barack Obama tours SpaceX with CEO Elon Musk at the Kennedy Space Center in Cape Canaveral, Florida, April 15, 2010.

Official White House Photo by Chuck Kennedy

in federal contracts by the end of 2008. One and a half billion dollars for the company that, just five years earlier, hauled its rocket on a truck across the country just to get the government to pay attention.

Even after this significant rearrangement, however, private companies were still primarily functioning as government contractors. As Boeing and Lockheed Martin had done since the dawn of the space age, new companies like SpaceX and Blue Origin vied to perform critical functions for NASA, still under the direction of a federal program. To accomplish the technological and commercial explosion Obama and his corporate partners envisioned, private companies would need to be able to follow their own pursuits in space—after all, it's hard to sell investors on the dream of carrying fresh fruit and clean laundry to the space station. What would be in it for the shareholders?

Prospects for the cosmic speculator seemed grim until November 2015, when a Republican-led House and Senate passed a remarkably bipartisan bill called the Commercial Space Launch Competitiveness Act.[42] Among other things, this legislation declares that anything "a U.S. citizen" manages to "recover" from an asteroid or planetary body can be owned, used, and sold by that US citizen. In other words, corporations (which, under US law, count as "persons") can extract resources from space and make money from them. Nascent space-mining companies like Planetary Resources and Moon Express rejoiced at the legislation, which opened up the heavens to industrial "development." Finally, there was a way to make money in space.

Although numerous legal scholars have questioned the right of the US to declare its citizens' ownership over extraterrestrial minerals, it is hard to imagine who could stop them. We will explore the thicket of international space law later in the book, but the upshot is that American companies are proceeding at breakneck speed to fill, drill, and populate outer space—and the United Nations will either catch up at some point or it won't.

Obama's astrocorporate breakthrough has indeed provided the shot in the arm he hoped to deliver to NASA. Just a few years later, Vice President Mike Pence took swipes at the previous adminis- tration while reaffirming its turn to the private sector, declaring to the newly reconstituted National Space Council that the US would return to the Moon and establish a mission to Mars "by any means necessary."[43] Ideally, Pence said, NASA would steer this project, but during the Obama administration, the agency had once more fallen into "the paralysis of analysis" while Russia and China continued rapidly to advance their own endeavors in space. It was impera- tive that the next humans on the Moon be "American astronauts launched by American astronauts on American soil," and if NASA couldn't make it happen, someone else would. "If commercial rockets are the only way to get American astronauts to the Moon in the next five years," Pence shrugged, "then commercial rockets it will be."

His message for NASA? "Step up." And if you don't, we'll just call Musk or Bezos or Branson. Because regardless of the means, he promised, *we're going to the Moon*. Again. (A day later, the *Washing- ton Post* ran an editorial titled "Mike Pence, Boldly Sending Amer- ica Back to Where Man Has Gone Before.")

"Let's go with confidence," Pence said, building to a predictably pious conclusion, "and let's go with faith." Squinting, nodding, try- ing to look sincere, Pence listed the objects of this faith: America, American ingenuity, American courage, the American pioneering spirit. "And lastly," he sighed, "let's, uh, let's have that other kind of faith as well." The kind of faith that assures us that the next bril- liant, brave Americans to go to space "will not go alone." At this point, Pence's prose gets downright purple:

For as millions of Americans have cherished throughout our long and storied history of exploration by this nation, let's believe, as the Old Book says, that there's nowhere we can go from His spirit. If we rise on the wings of the dawn, if we settle on the far side of the sea, *even if we go up to the heavens*, there His hand will guide us [applause]. And His right hand will hold us fast [more applause].

I'm going to take some time with this passage because it so clearly uses religion to bind together a tangle of secular aspirations. Just in case the military, technological, and economic arguments don't work, Pence is suggesting, here's some divine assurance that we're doing the right thing.

First, I should point out that there's no such thing as "the Old Book." Pence is referring to the Bible, which some Christians call "the Good Book," a nickname he seems to have crossed with "the Old Testament." Pence is clearly going for an expression of familiarity here, but he ends up sounding like he doesn't know what he's talking about. No one calls it "the Old Book"—no one, that is, except a few nineteenth-century revivalists seeking to make America Christian "again."[44] (So maybe that's what Pence is up to, after all.)

The lofty language about God's guiding us no matter where we go is from Psalm 139, which professes faith in God's omnipresence— even his inescapability. Wherever I go, for better or for worse, I can't get away from this God. Lying down, rising up, on the land or the sea or in heaven or hell, says the psalmist, "there your hand shall lead me, and your right hand shall hold me fast."[45] When Pence invokes the "millions of Americans" who've been guided by this passage, he reactivates divine sanction for American frontierism. Just as God directed the pilgrims to build a holy land in the "New World," just as he directed their descendants to extend this land out West, God is now calling Americans to a new frontier out in space.

A year after Pence's address to the National Space Council, President Donald Trump stirred up a minor journalistic and academic outcry by calling our future in the stars America's "Manifest Destiny."[46] A product of the nineteenth-century westward expansion, this doctrine justified the seizure of Indigenous land by calling it the will of God (perhaps as communicated in "the Old Book"). And although Pence himself seems reluctant to use the word *manifest*, he is clearly invoking the idea by channeling Psalm 139. In fact, he's extending the old doctrine infinitely by applying it to the endless expanse of space. Just as God was with "us" when "we" crossed the seas, battled the Red Coats, forded the rivers, and mined the mountains for gold, God will be there "even if we go up to the *heavens.*"

Granted, Trump's and Pence's appeals to God are more rhetorical than devotional. They appeal to Trump's Evangelical Christian base while filling secular listeners with an inchoate feeling of somber significance. Most of the engineers at NASA, SpaceX, and Blue Origin, along with their online devotees, would shudder at the idea that their work is in any way religious. And yet, as Trump and Pence make clear, the idea of a human *mission* to colonize unknown lands has a specifically religious history that still animates the contemporary space program and justifies our claim to whatever we lay our probes on.

Even if we mine the Moon, commodify space, and live in a Club Med-ded tin can; even if we don't *believe* in the God who allegedly ordains the whole affair, it's "the Old Book" that assures us that the universe is ours. So it's back to the Bible we go.

CHAPTER TWO

Creation and Conquest

The conquest of the earth . . . is not a pretty thing when you look into it too much.
JOSEPH CONRAD

Inheritance

When my great-grandmother was no longer able to host Passover Seders, she skipped right over her cranky atheist son and gave the Haggadahs (prayer books) to her grandson instead. A more cheerful agnostic who happens to be named Joshua and happens to be my father, he promised he'd conduct the annual service with our family's characteristic blend of irreverence, gratitude, and critical interjections.

Passover commemorates God's deliverance of the Israelites out of slavery into freedom. It tells the story of Moses parting the Red Sea so the Israelites could flee Egypt, enter the wilderness of Sinai, receive the Ten Commandments, and cross the Jordan with Joshua into the land of Canaan, the "land flowing with milk and honey" that God promised to the descendants of Abraham (Exodus 3:17).[1]

"Next year at this season, may the whole house of Israel be free!" the Seder leader says before inviting the youngest child to recite the Four Questions. "And the whole house of Palestine, too," one of my siblings will interject, having already interrupted the service to mourn the Egyptians whom God senselessly drowns in the sea.

"And Tibet and Venezuela," another will chime in, "and refugees everywhere."

"And may the police stop killing Black people."

"Amen."

Eventually, we'll proceed to the Four Questions, which someone else will interrupt by retelling the story of my grandfather's having taught my four-year-old father the wrong text for his first recitation. Rising to his little feet under the devout eyes of the old folks, my father did *not* ask, "Why is this night different from all other nights," because those weren't the words my grandfather taught him. Instead, this mini-Joshua said, "Religion is the opiate of the people. You have nothing to lose but your chains."

"Why the hell would I come to the Seder?" my grandfather would grumble on the phone to me in the early spring. "Why should I care where a bunch of people wandered thousands of years ago when they decided to make up a god?" I never had a decent answer for him ("Umm . . . the witty repartee?"), but he'd always come anyway, looking annoyed in his corner chair and then perking up when someone diverted the conversation to crooked politicians, universal health care, or Serena Williams.

Whatever your temperament, you may find yourself wondering along with my late grandfather why you should care about Bible stories in a book about outer space. After all, the Bible is ancient history, whereas space is our aspirational future. The Bible revolves around faith and ritual, whereas space occupies the realm of science and technology. What are we even doing here?

* * *

In the last chapter, we heard both Mike Pence and Donald Trump claim divine sanction for the specifically American seizure of outer space. This strategy—of justifying political domination through religious rhetoric—has a long, unsavory history in the Americas in particular. As biblical scholar Michael Prior explains, "colonizers invariably seek out some ideological principle to justify their actions," and the more violent and exploitative the endeavor, the loftier the ideology.[2] From the conquest of the "New World" to the establishment of South African apartheid to the Cold War Space Race, the project of empire building usually claims "altruistic motives" and often insists it's advancing a sacred destiny.[3]

One might recall Elon Musk defending what Bernie Sanders called his "obscene" wealth by saying he's "extending the light of consciousness to the stars." Or Jeff Bezos extolling the future he's building for his grandchildren and Mark Zuckerberg's intellectual heirs. These grand, seemingly selfless motivations hide the territorialism, economic injustice, military collusion, scientific cooptation, and environmental recklessness that actually fuel NewSpace. Anyone who dares in the company of believers to contest this vision (I've tried) tends to be silenced with pious incantations about the salvation of humanity. And although the contemporary spacenik crowd is more easily stirred by the sci-fi canon than the Old or New Testaments, the group of writings we have come to call "the Bible" nevertheless establishes the major principles that produce and sustain the NewSpace race. So this is going to sound grandiose, but I'll stand by it: the reason it's important to head back to the Bible is that we can't understand Western politics, Western ethics, or Western science without it, and it's these arenas that justify the otherworldly mission of NewSpace.

The Sacred Roots of Secular Culture

The notion that Western secularism is a direct product of Western religion is at least as old as Friedrich Nietzsche (1844–1900). Best known for his cryptic proclamation that "God is dead," Nietzsche spent his increasingly moody authorship trying to release the stranglehold Judaism and Christianity held on the modern world and to let different values emerge. For Nietzsche, the problem was primarily psychological: the guilt-inducing, pleasure-denying monotheisms have shaped us into human beings who, whether religious or not, can't stand ourselves. Thanks to the global, nearly total spread of "Jewish" ideals through Christian empire, even those of us who think we are "free, *very* free spirits" actively try to deny everything in us that's strong, healthy, and life affirming.[4] In short, Nietzsche charged, Christianity has made us into *nihilists*.

The culprit, he claimed, is the doctrine of sin. According to the specific kind of Christianity that teamed up with European imperialism, all of creation is "fallen" and filled with temptations. Given the mess we've made of the world, the best we can do is detach ourselves from earthly pleasures and suffer to make ourselves worthy of heaven, where we will finally be happy, healthy, and free (can you hear the similarities to Musk's Martian aspirations?).

It will probably come as no surprise that Nietzsche didn't think there *was* a heaven. Far from existing somewhere "out there," heaven was an idea that priests made up to keep themselves in power and keep the masses calm. "Oh this insane, pathetic beast—man,"[5] Nietzsche fumed. Under the thumb of a God that "man" himself made up, this "pathetic beast" denies the importance of this world—*this whole world*—to wait for another world that doesn't even exist. This is what Nietzsche meant when he called Christians *nihilists* (from the Latin word for "nothing"): they don't believe in the world that does exist and do believe in a world that doesn't. So they actually believe in *nothing*.

As Nietzsche argues, this religious will-toward-nothing has produced a nihilistic *politics* (think of all those utopias that never arrive), a nihilistic *ethics* (consider the minimalist "charity" that keeps rich folks rich and poor folks poor), a nihilistic *culture* (as in those self-flagellating exercise, diet, and skincare regimens that only make us hate ourselves more), and, as its crowning achievement, a nihilistic *science*.

He knows it sounds weird. Religion produces *science*? It's a question I always get in taxis and airplanes when I tell people what I teach and write about. "What do you mean, science and religion?" my new friend will stammer. "Aren't those opposites?"

"No!" Nietzsche thunders back. "Don't come to me with science when I ask for the natural antagonist to [religion]."[6] One can almost hear the man snuffing out air and stamping his foot. For better or worse, I try to frame it more politely than Nietzsche does, but the point is that secular science is just as world-denying as its sacred ancestor. Think about all those monkish hours scientists spend in the lab, or behind the computer screen, or under the telescope—searching just as ardently as pilgrims for *truth*. A truth, moreover, that they believe not to be created or subjective but rather eternal and universal.

Science as a direct descendant of religion. The more you think about it, the less weird it seems. The universal claims. The exhausting ritual practice. The functional priesthood of scientists themselves, who speak a language no one else understands, who claim direct access to the truth, and who then translate and disseminate that truth to the rest of us. The physicians who do the diagnostic and healing work that other cultures assign to shamans, exorcists, rabbis, mediums, and ministers. The grandiose myths of the big bang and natural selection that seek to replace our stories of creative gods and chaos monsters.

Seen in this light, science does seem to have inherited many of the values and functions of religion. The question is whether this genealogy matters. And the reason I'd like to suggest it does is that the very nihilism that characterized imperial Christianity now characterizes modern science—at least those forms of it that enthusiastically advance military and financial pursuits. In short, the institutional sciences have taken up imperial Christianity's world-denying, even world-destroying ideology. An ideology that insists it's not an ideology but just *the truth*.

When the former CPO of Tinder recently tweeted that the smartest people he knew had "physics backgrounds," Musk responded that "Physics is simply the search for truth. Nothing is more rigorous."[7] So the situation hasn't changed much since Nietzsche's time. Just like the Western monotheisms, Western "science" believes in a single truth—not created, not invented, but somewhere "out there," waiting to be discovered and believed.

* * *

The clearest illustration of the religious inheritance of modern science is the escalating environmental disaster. That, at least, was the conviction of a historian named Lynn White, who, thanks to the work of Rachel Carson and other early ecologists, could see by the mid-1960s that the planet was in serious danger. Something needed to be done to reverse the course of deforestation, mass extinction, and pollution on which we'd set our Earth. But, White maintained, this "something" could not just be more science and technology, because science and technology had created the problem in the first place. The key, he insisted, was to "rethink our axioms" by figuring out where they came from.[8] They key was to unearth "the historical roots of our ecologic crisis" so we might plant ourselves in different soil.

White digs up these roots in quick, backward stages. Today's environmental crisis, he says, can be traced back to the nineteenth-century "marriage between science and technology," which equated the *knowledge of nature* with the *power to manipulate it*.[9] Of course, he adds, the West was already manipulating nature by the late twelfth century, when giant mills harnessed the power of wind. The West was even manipulating nature in the *seventh* century, when the scratch plows that gently tilled the earth gave way to wheeled plows that "attacked" it. So according to White, the real turning point in the development of Western techno-science took place before all these developments, with "the victory of Christianity over paganism."[10]

What exactly is White talking about? What does the Christian victory over paganism have to do with the techno-science that's destroying our planet?

First, I should say that *paganism* is a particularly unhelpful word. Until the nineteenth century, in fact, people didn't even call themselves pagans. Rather, "paganism" was an "all embracing, pejorative term" to refer to non-Jews and non-Christians.[11] In that sense, it was akin to "heathenism" or "idolatry": all people who worshipped deities other than the God of Israel were gathered under the blanket denigration of *pagan*, from the Latin word meaning "ordinary person," or "commoner." So for most of the word's history, *paganism* designated teachings and practices that monotheists deemed unsophisticated and untrue.

In the New Testament's Gospel of Matthew, the risen Jesus tells his disciples to "go . . . and make disciples of all nations, baptizing them in the name of the Father and of the Son and of the Holy Spirit" (Matthew 28:19). Especially once this project of Christian disciple-ship became entangled with Roman imperialism, the instruction served as an ideological justification for converting or destroy-ing "pagans" wherever they might be found. Thus the gods of the Anglo-Saxons, Eastern Europeans, Greeks and Romans, North Africans, sub-Saharan Africans, Aboriginal Australians, North Americans, South Americans, North Indians, South Indians—you get the idea—were systematically unseated by the God of Abraham, Isaac, and Jacob. Particularly by this God as he'd revealed himself in Jesus of Nazareth, filtered himself through the Apostle Paul, and joined forces with terrestrial empire.

But, again, what does this Christian ousting of "paganism" have to do with modern technology and ecological destruction?

Although European, American, Australian, African, and Indian "pagans" have produced vastly different ethics and mythologies, early-modern scholars detected in most a tendency that European anthropologists named *animism*.[12] From the Latin word for "spirit" or "soul," animism names a way of relating to the "things" of the natural world—like forests, rivers, rocks, and trees—as living, sen-tient, personal beings, some of whom are sacred and even divine. In the Anglo-Saxon world, for example, "before one cut a tree, mined a mountain, or dammed a brook, it was important to placate the spirit in charge of that particular situation, and to keep it pla-cated."[13] With the incursion of Christianity, however, these more-than-human beings became inanimate. In the Christian worldview, there *were* no tree people or river spirits, so there was no need to appeal to or appease them before changing, using, or abusing them. "By destroying pagan animism," White explains, "Christianity made it possible to exploit nature in a mood of indifference to the feelings of natural objects."[14]

We might note here White's use of the word *objects*. What he's arguing is that the Christian victory over paganism turned a world full of people into a world full of things, and it's here that we can finally see the connection to the ecological crisis. Given its equation of *knowledge of* nature with *power over* nature, Western techno-science needs to view the natural world as composed of objects rather than subjects. After all, if the mountaintops were ancestors, then we couldn't remove them. If the stones were persons, we couldn't frack them. If the forests had spirits, we'd hesitate to clear-cut them. If the rivers were sacred, we wouldn't fill them with radioactive waste. If rats and mice were kin, we wouldn't test our toxic cosmetics on them. And if the land itself were sacred, we probably wouldn't overfarm it within an inch of its life. We'd work with it rather than against it, give it rest, only take what we need, and give back whatever we could. But thanks to the victory of imperial Christianity, Western communities and institutions don't have these sorts of relationships with the land. Thanks to the victory of imperial Christianity, the land isn't a person we might relate to in the first place.

And again, White argues that it was this despiriting of nature that enabled the emergence of a distinctively Western techno-science. A way of understanding and manipulating the world as a set of resources for human "progress," wealth, and comfort. A way of understanding and manipulating the world that's rapidly bringing an end to our world.

Having just sketched the eco-destructive legacy of Christianity, and before I go on to fill in the picture, I'd like to take a moment to make a massive qualification. With the exception of some particularly vocal American conservative Evangelicals,[15] most contemporary Christian texts, leaders, and communities are solidly committed to environmental stewardship. Thanks in no small part to Lynn White's call either to "find a new religion, or rethink our old one," almost all contemporary theology could be called ecotheology; nearly every text in this field puts environmental justice and repair among its most central concerns.[16] Most major denominations have issued statements on the sacred mandate to care for creation, priests and ministers give impassioned Earth Day sermons each April, and major cathedrals celebrate the interspecies ministry of St. Francis each October.[17] Perhaps most famously, the current Roman Catholic pope, who shocked the world by choosing St. Francis as his namesake, wrote an unprecedented environmental encyclical that condemned contemporary capitalism's vision of humanity as "lords and masters," of "our Sister, Mother Earth . . . entitled to plunder her at will."[18] And as hurricanes and fires roared through the summer of 2021, the leaders of the three largest Christian denominations issued "A Joint Statement for the Protection of Creation," calling on "everyone, whatever their belief or worldview, to . . . listen to the cry of the earth . . . , pledging meaningful sacrifices for the sake of the earth which God has given us."[19]

To put it bluntly, anybody who's anybody in Christian circles today is deeply concerned about the state of the oceans, the species, and the climate. And many of these voices are courageous enough to demand that their churches take responsibility for Christianity's complicity in Western ecocide. So Ecumenical Patriarch Bartholomew calls the Eastern Orthodox Church "to acknowledge 'our contribution . . . to the disfigurement and destruction of creation'"; the Evangelical Lutheran Church in America confesses that "we [have] become captives to demonic powers and unjust institutions . . . [that] treat the earth as a boundless warehouse"; and Pope Francis admits that "we Christians have at times incorrectly interpreted the Scriptures" as justifying "the unbridled exploitation of nature."[20] These leaders all agree not only that individual humans must become good stewards of the God-given Earth (by recycling, planting trees, shopping less, etc.) but also that corporate and governmental institutions must overhaul their practices—beginning with the untrammeled pursuit of profit.[21]

The limitation of this nearly unanimous ecclesiastical outpouring is that, American Christian nationalism notwithstanding, the churches wield far less influence in the contemporary world than they did during the medieval and early-modern periods. In short, the churches don't have nearly as much power to fix our ecological crisis as they did to produce it. What nation or corporation would be inclined at this point to heed the chastened counsel of social-justicey Christians? Is SpaceX really going to consult a papal encyclical before it crashes another rocket into a threatened coral reef?

The problem is that the very doctrines the churches now condemn as bad readings of Scripture (the supremacy of humanity, the exploitation of the Earth, the inanimacy of the land) have now installed themselves at the heart of the imperial politics, capitalist economics, and secular science that animate the astrotopian dream of escaping Earth. What began as biblical teachings have become so commonplace that they now masquerade as secular, universal, and therefore unchangeable, principles. And chief among these biblical-turned-secular principles is the "dominion" of human beings over the rest of creation.

Fill the Earth and Subdue It

One of the particularly destructive myths guiding the contemporary space race is the puzzling idea that humans are the most important thing around. If you take a minute, you'll find it everywhere: the US and Soviet space agencies sent dogs and chimps into orbit before sending human beings; pharmaceutical companies test chemicals on mice before they ever test them on people; elk and caribou have no say in zoning meetings about new highways; meadows and rainforests have no legal standing—we could go on for days.

Here comes another big qualification, this time for the scientists and citizens of the modern world. I am fully aware that individual *people* maintain all sorts of respectful, loving relationships with the more-than-human world. This or that natural or social scientist, myself included, most likely does the work she does out of awe and even reverence for the animals, plants, bacteria, and minerals that surround and compose us. But almost all of us work for, and rely on, *institutions*. And nearly every modern Western institution, from the university to the military to the private sector, operates under the unexamined assumption that the natural world belongs to humanity to use as we see fit. There are numerous sources of this unexamined assumption—capitalism, colonialism, extractivism, and so forth—but if we follow Lynn White's lead, we'll land all the way back in the Bible.

The first chapter of the first book of the Old Testament asserts the supremacy of humanity over everything else in the world. Although it calls each stage of creation "good," the opening creation myth in Genesis confers "the image of God" on human beings alone (Genesis 1:26–27). This status bestows on humanity powers analogous to the powers of God: in the very next verse, God tells the humans to "be fruitful and multiply, and fill the earth and subdue it." In other words, humans are directed to take on the divine power of creation and to rule the created world from within. In case the reader is uncertain of the scope of such filling and subduing, God gets more specific, telling his humans to "have dominion over the fish of the sea and over the birds of the air and over every living thing that moves upon the earth" (Genesis 1:28). Because human beings are made "in the image and likeness of God," they receive from him the godly powers of reproduction and dominion, of creation and conquest.

As it turns out, the next chapter of Genesis offers a totally different creation story from the first. Contemporary biblical scholars agree that these two stories were actually composed hundreds of years apart and were later stitched together by some editor's half-hearted, transitional verses.[22] Even a casual reading of the two myths reveals striking differences between them. If Genesis 1 gives us a disembodied creator speaking the world into being ("Let there be light," "Let the waters bring forth swarms of living creatures," etc.), Genesis 2 brings God down to Earth, tinkering and sculpting. As the scene opens, we find God shaping an Earth creature out of dust and breathing life into his nostrils. The medium may be different, but the message seems the same: the human is the crown of creation. God grows a garden *for him*, makes a slew of animals *for him*, and gives him the power to name every other creature in the land and sea. And when, after all that, the human being is still lonely, God makes him a girlfriend.

At this point, we've come upon the most notable difference between these two creation stories. While Genesis 1 always speaks of "male and female" in the same breath, Genesis 2 differentiates them—temporally, temperamentally, and functionally. It's Genesis 2 that gives us that genuinely strange story of God's shaping Eve out of Adam's rib (strange because even though our experience tells us that every human being has been born from the midsection of a woman, Genesis insists that the first woman was born from the midsection of a *man*). It's Genesis 2 that tells us God made the woman "for" the man, and it's Genesis 2 that records the man's naming the woman, just as he did the nonhuman animals. Considering that Eve was created both for and out of the man, the Apostle Paul concludes that "man" alone "is the image and reflection of God" while "woman is the reflection of man" (1 Corinthians 11:8–9), a step above the animals and a step below her husband.

Taken together, the two creation stories of Genesis—especially as filtered through Christian readers of Paul—leave us with the broad sense that humans are subject to God, women are subject to men, and the entirety of the natural world is subject to humans.[23] As the biblical narrative moves from creation to history, however, this hierarchy becomes even less egalitarian. With the coming of the covenant, we learn not only that male humans are superior to female humans but also that the people of God are superior to the people of every other nation.

The Canaan Complex

When I first listened to Mike Pence's address to the National Space Council (the one where he orders America back to the Moon with American astronauts on American rockets from American soil), I was absolutely horrified by his recitation of Psalm 139. The one that assures us that, wherever we are and wherever we go, God will be with us. When Pence says that "even if we go up to the heavens, there [God's] hand will guide us," he is claiming divine sanction for the American military adventure in outer space. As we will see in the next two chapters, this is a familiar move in US history. The same God who "gave" Europe the New World and pushed white settlers across the continent is now allegedly granting the US the Moon, Mars, the asteroid belt, and anything else it manages to land on.

"It's like an infinite Canaan," I said to my sister, fuming as I chopped the potatoes for dinner that evening. "Everything we see is ours. *This* is the problem with the new corporate space race," I carried on (*chop, chop*); "America's still got a Canaan complex." My sister blinked at me. "Ooh!" I shouted, dropping the knife; "there's my book title! *The Canaan Complex*. Wait, no; *Infinite Canaan*. Or how about *Cosmic Canaan*? What do you think? Which one's best?"

"Um," my sister responded, making one of those teeth-grippy faces, "am I supposed to know this? I am, right? So sorry—what's Canaan again?" This from the Rabbi's pet in Hebrew school. Okay, I'd have to find a new title.

Geographically speaking, the land the Hebrew Bible calls Canaan lies east of the Mediterranean Sea and encompasses modern-day Jordan, Israel, Gaza, the West Bank, and the southern parts of Syria and Lebanon. Mythically speaking, Canaan is "the Promised Land"—the land "flowing with milk and honey" (Exodus 3:17) that God pledges to an otherwise ordinary guy named Abram for reasons the text doesn't give us.

In two separate scenes, God promises two things to Abram: children and land. "Look toward heaven and count the stars," God tells Abram; "so shall your descendants be" (Genesis 15:5). These innumerable offspring will eventually live in Canaan, which God now promises to Abram, saying,

To your descendants I give this land, from the river of Egypt to the great river, the river Euphrates, the land of the Kenites, the Kenizzites, the Kadmonites, the Hittites, the Perizzites, the Rephaim, the Amorites, the Canaanites, the Girgashites, and the Jebusites. (Genesis 15:18–21)

When I visualize this scene, I tend to break it into a split screen. On the left side, Abram is looking up at the stars. On the right, he's looking out at the land. Up and out: even in the biblical literature, the infinite stars promise earthly prosperity.

As you may have heard, however, such prosperity doesn't come easily—or quickly. God does warn Abram it's going to take a while to fulfill the Covenant, and the ensuing six books of the Bible narrate the long travails of the people of the promise on their way to inhabiting the land. Abram and Sarai are renamed Abraham and Sarah. They circumcise their relatives. They're a hundred years old and still childless. God finally gives them a son, Isaac, and then tells Abraham to sacrifice him. God saves Isaac with a ram, chooses Jacob over Esau, and lets Joseph get sold into slavery. God frees Joseph but then tells his whole family to move to Egypt, where they'll all become slaves. And then "with a mighty hand," God sends the liberator Moses, kills the firstborn sons of the Egyptians, and drowns even more of them in the Red Sea. God then lets the Israelites wander in the desert for forty years, guides them in a pillar of cloud by day and a pillar of fire by night, feeds them with manna from the heavens, gives them the Ten Commandments (and hundreds of others), tells Moses that after all that, the poor guy's not actually going to make it to Canaan, and finally appoints Joshua to deliver the people across the Jordan and into the Promised Land.

But here's where things get really troubling. There are people in the Promised Land. Remember that long list of names? "The Kenites, the Kenizzites, the Kadmonites, the Hittites, the Perizzites, the Rephaim, the Amorites, the Canaanites, the Girgashites, and the Jebusites"? Presumably, these communities believe the land belongs to them—and that they belong to it. How will the people of God convince them it belongs to the Israelites? Rational argument? Rousing sermons? Political treaties?

"When you cross over the Jordan into the land of Canaan," God says to Moses, "you shall drive out all the inhabitants of the land from before you, destroy all their figured stones, destroy all their cast images, and demolish all their high places" (Numbers 33:51–

52). In other words, you will get rid of the people, their sacred things, and their sacred spaces. As God goes on to explain, if the Israelites don't clear out everyone and everything, then the Indigenous remnant might seduce the Israelites into idolatry (what the Christians will come to call paganism). Therefore, he demands in Deuteronomy, "you must utterly destroy them. Make no covenant with them and show them no mercy. Do not intermarry with them . . . for that would turn away your children from following me, to serve other gods" (Deuteronomy 7:2–4). And again, the key to destroying these idolatrous people is to destroy their idolatrous world: "Break down their altars, smash their pillars, hew down their sacred poles, and burn their idols with fire" (Deuteronomy 7:5).

In these passages, the obliteration of the inhabitants of Canaan is not only divinely endorsed but divinely mandated. If the Israelites fail to destroy the Kenites, the Kenizzites, the Kadmonites, and their neighbors, then God will abandon and curse them, "for the Lord your God is a devouring fire, a jealous God" (Deuteronomy 28:15–68, 4:24). But if the Israelites "devour all the peoples . . . showing them no pity" (Deuteronomy 7:16), then God will reward them, giving them children, cattle, grain, wine, oil, and, incidentally, iron and copper to mine from the newly empty hills and stones (Deuteronomy 8:9). So devouring the peoples it is.

Exodus. The deliverance from slavery, the parting of the Red Sea, the drowning of Pharaoh's army, and the Lord's constant presence as a pillar of cloud by day and of fire by night. Then finally, after the forty years of wandering, the giving of the law, the promises and threats, and the anticlimactic death of Moses, God tells Joshua it's time to conquer Canaan. "So Joshua defeated the whole land, the hill country and the Negeb and the lowland and the slopes, and all their kings; he left no one remaining, but *utterly destroyed all that breathed*, as the Lord God of Israel commanded" (Joshua 10:40).

There are more passages, just as horrible as the ones I've cited here, which ring over and over like a deranged alarm clock as the conquest approaches. They even echo periodically after it's all over.[24] The worst part is that these murderous passages make the benign and even loving ones seem sinister. Those verses that one actually hears in synagogue or church about liberation, caring for the widowed and orphaned, and God's tireless presence with God's people. When you sit down and read the whole text, these beautiful verses are drowned out and even completely recoded by the genocidal refrains that punctuate them.

Another massive caveat here: the conquest of Canaan didn't actually happen. There's simply no evidence that the Canaanites were annihilated by, expelled by, or even ethnically distinct from the Israelites at all.[25] In fact, the story was written during and after the Babylonian exile (ca. 597–539 BCE), centuries after the Exodus and Eisodus (or entrance into the Promised Land) would have taken place. So rather than being a historical account of military victory, the conquest of Canaan is a retrospective fantasy written by a displaced and dominated people—a people supposedly under the protection of a powerful God—wondering what the hell they did wrong.

Unfortunately, debunking the history does nothing to displace the story itself. Even if there was, in fact, no such thing as the conquest of Canaan, we still have what Osage literary scholar Robert Allen Warrior calls a "narrative problem," which is that the story has gone on to justify the expulsion and extermination of Indigenous people in North America, Ireland, Australia, South Africa, Eastern Europe, and Palestine alike.[26] So although we might breathe a bit more easily knowing that the Canaanites were not actually obliterated by the Israelites, the relief doesn't last long considering the systematic exterminations the text has nevertheless underwritten. Hence my family's squirmy revisions of the Seder. Hence, perhaps, my energetically bat-mitzvahed sister forgetting what Canaan was.

Just before he heads in to destroy "everything that breathes," Joshua pauses to listen to God, who assures his servant, "The Lord your God is with you wherever you go" (Joshua 1:9). Taken on its own, this is a beautiful, emboldening promise: no matter the challenge or the danger, God is with us. But in the context of the ruthless slaughter of all sentient beings, whether or not it actually happened, the promise of divine presence becomes downright terrifying. To say that God is "with" Joshua is to sanctify the annihilation of every foreign person and thing he encounters—especially whatever the Canaanite Natives hold sacred.

This is the force behind the "God" who backs European conquerors and American politicians—the nationalist lineage that produces Pence's intention to grab hold of outer space while God's right hand holds him fast. From Canaan through the Americas and even into the heavens, to say that God is "with" the invaders is to claim not only divine approval for the conquest but a divine mandate for it. And even if no one believes it, declaring a divine mandate to the land has historically been enough to make it so. Especially in the Americas.

CHAPTER THREE

The American Promised Land

Gentlemen! It is the War of the Lord which you are now Engaged in.
COTTON MATHER

Sacred Symbols of a Secular Nation

"The turkey is a truly noble bird." It's by far the best line from 1776, the musical about the American founding fathers that folks haven't mentioned much since the more ethically nuanced and compositionally sophisticated *Hamilton* left theatergoers dazed at the end of the Obama era. These days, I hear that high school students watch Lin-Manuel Miranda and Daveed Diggs light up Disney+ in their US history classes. For us, it was Howard da Silva on a worn VHS, imploring William Daniels's John Adams to "sit down" while rejecting his choice of the national bird.

> **Franklin**. What sort of bird shall we choose as the symbol of our new America?
>
> **Adams**. The eagle.
>
> **Jefferson**. The dove.
>
> **Franklin**. The turkey.
>
> . . . .
>
> **Adams**. The eagle is a majestic bird!
>
> **Franklin**. The eagle is a scavenger, a thief, and a coward. A symbol of over ten centuries of European mischief.

Adams. The turkey?

Franklin. The turkey is a truly noble bird.

The turkey's nobility, says Franklin, lies in its being "Native American. Source of sustenance of our original settlers. An incredibly brave fellow, who will not flinch at attacking a regiment of Englishmen, single-handedly."[1]

As you've no doubt heard, however, Adams prevailed in this matter, establishing the bald eagle as the sacred symbol of his secular America. The eagle adorns the Great Seal of the United States, while the turkey is doomed to adorn dinner tables every year at Thanksgiving. The sacred and the sacrificed.

To call something "sacred" is to set it apart from the ordinary world. A sacred thing is a special thing, and as such, it must be treated with care and respect. Think of the special rules and prohibitions around approaching an altar, entering a national park, or flying on a plane with a pop star. The things we hold sacred are the things we most energetically protect. If Franklin's satirically proposed turkey had become our national bird, it would have become just as unthinkable to eat one of them as it is now to eat an eagle. But the fearless, "Native American" turkey lost out to the cowardly eagle, whom Franklin called "a bird of bad moral character" for stealing the fish out of other birds' mouths.[2]

The national bird was not the only national symbol Franklin failed to secure in the early days of the Republic. The other one, uncommemorated so far as I know in song or dance, was the Great Seal itself. At their first meeting to determine its design, Franklin, Jefferson, and Adams all agreed the seal should depict not a creature but a story. The Great Seal of the United States should, in their minds, imprint on the nation a collective myth of origin to steer it toward a shared future.

John Adams proposed an engraving of "The Judgment of Hercules," in which the hero is tempted by "Vice" but decides on "Virtue." Forsaking all the "flowery Paths of Pleasure" revealed to him by a beautiful, recumbent Sloth, Adams's Hercules opts for the "rugged

Mountain" of hard work and self-control.[3] The future president's message was clear: the bounty of the land could lead America into decadent decline or austere elevation, and Adams was commending the latter. For their part, however, Franklin and Jefferson turned from Greek mythology to the Hebrew Bible, each proposing that the Great Seal feature a scene from the book of Exodus.

Franklin suggested the crossing of the Red Sea. Specifically, he wanted to commemorate that moment when the Israelites made it to dry land and Moses raised his staff to close the partition. Pharaoh, his army, their chariots, and horses are all drowned, and the people of God rejoice. In case any contemporary viewer might be troubled by this deadly scene, Franklin specified that there should be "rays from a Pillar of Fire in the Clouds reaching to Moses, to express that he acts by Command of the Deity." To drive the point home, Franklin proposed that the national motto be "Rebellion to Tyrants is Obedience to God."[4] Jefferson endorsed Franklin's motto

Figure 3 . 1 1856 rendering of Benjamin Franklin's proposed Great Seal of the United States.

Department of State

Figure 3.2 1903 rendering of Thomas Jefferson's proposed Great Seal of the United States.

Department of State

but scrolled to a later part of the Exodus story, proposing an image of the Israelites in the wilderness guided by God in a pillar of cloud by day and a pillar of fire by night.

It may seem perplexing that Benjamin Franklin and Thomas Jefferson, whose religious sensibilities were muffled at best, each sought to map the American narrative onto the Israelite one. Or it might seem an extraordinary coincidence: of the thousands of stories in the Bible, *this* is the one they both chose? As it turns out, however, Franklin and Jefferson were merely echoing the previous century's incessant refrain: America was "God's New Israel," her people delivered from bondage in the Old World to a sparkling New Canaan across the seas.[5] This is the analogy Mike Pence will extend into the heavens, claiming God's blessing as America storms the Infinite Promised Land.

The sources of this sacralized American bravado are frankly innumerable, ranging from chaplains and explorers to revivalists, university presidents, governors, journalists, and presidents. And the theme is constant: "America has been elected by God for a special destiny in the world."[6] Over and over, as if to convince themselves it's true, early American sermons, letters, and speeches sketch out and fill in the analogy: England is the oppressive Egypt; Charles I is the obstinate Pharaoh; the Puritans and planters are the chosen people; the Atlantic journey is the crossing of the Red Sea; North America is the Promised Land; George Washington is Moses, Joshua, or both; and the United States will be an American Israel, a "light to the nations" in the way of freedom, benevolence, and peace.

But just as Moses and Joshua knew that the land of Canaan was already occupied, the European settlers, colonists, and founding fathers knew there were already people in "their" Promised Land. So they completed the analogy. If America was the New Israel, then Indigenous Americans must be the Amorites, Hittites, and Jebusites—that is, the "Canaanites" whom God told the prophets to remove and destroy. As one colonial official wrote to King Phillip II of Spain in 1557,

> [America] is the Land of Promise, possessed by idolaters, the Amorite, Amalekite, Moabite, Canaanite. This is the land promised by the Eternal Father to the faithful, since we are commanded by God in the Holy Scriptures to take it from them, being idolaters, and, by reason of their idolatry and sin, to put them all to the knife, leaving no thing save maidens and children, their cities robbed and sacked, their walls and houses levelled to the earth.[7]

In short, comparing Indigenous Americans to the Canaanites lent divine justification to acts of violence that would be otherwise unthinkable.

In some sources, this comparison is allegorical: like the Canaanites, Indigenous Americans are said to be "idolaters," worshipping gods other than the God of Abraham, Isaac, and Jacob. Like the Canaanites, they are likened to beasts and insects and accused of vague sexual indiscretions that allegedly confirm their depravity.[8] In other sources, these similarities are taken as evidence of the Natives being *actual* Canaanites. As Yale University president Ezra Stiles explained in 1783, "the American *Indians*" were the literal descendants of the Canaanites whom Joshua drove out of Zion. Fleeing the army of God, the cursed ancient race climbed aboard Phoenician ships and sailed across the Atlantic to the "New World," a land from which God's New Israel would now further expel them.

Of course, one major difference between European settlers and the ancient Israelites was that the settlers were Christian. So although they used the Hebrew scriptures to justify their claims to the land and abuse of its caretakers, they also filtered those texts through specifically Christian doctrine, imagery, and myth. For example, Indigenous Americans were not merely "idolaters," as Deuteronomy would call them, but also "heathen," "pagans," "hellish fiends," "slaves of the devil," and "a generation of vipers even of Satan's own brood."[9] The land was not only Joshua's Canaan but also St. John's New Jerusalem: the "new heaven and new earth" that emerges out of the ashes of the Christian apocalypse (Revelation 21:1).[10] And perhaps most notably, under the Christian paradigm, there was an alternative to driving out the "Canaanites" on the one hand and destroying them on the other: the invaders could try to convert them. But as one Spanish legal theorist asked in 1544, how are we to convert the "barbarians" without conquering them first?[11] Even for early-modern "Christian Europe," what really mattered was possession of the land.

This Land Is *Whose* Land?

In Rio Vista, California, there is a man who will sell you the Moon. All you have to do is call or visit his Lunar Embassy website, specify the number of acres you'd like to buy, give him your credit card or PayPal details, and it's yours. A few weeks later, you'll receive confirmation of your property's coordinates and an official deed, printed on legal-size "simulation parchment paper suitable for framing."[12] You can also buy parts of Mars, Mercury, Venus, Jupiter's moon Io, and the entirety of Pluto.

Back in 1980, as the website recounts, "a very bright, young, and handsome" Mr. Dennis M. Hope wrote to the UN General Assembly, the American president, and the "Russian Government" to inform them he was asserting his right to the Moon at a claims office in San Francisco, a move he understood not to be prohibited by international law.[13] According to Hope, neither the UN nor the US nor the USSR ever responded. With no one trying to stop him, he subsequently laid claim to other moons and planets. Now Hope says he owns most of the solar system, which he parcels out to any earthling with twenty-five dollars and either a taste for novelty gifts or a genuine aspiration to own a piece of cosmic real estate.

For what it's worth, Dennis Hope doesn't seem to be a con man. He seems genuinely convinced that he's found a legal loophole wide enough to claim our cosmic neighborhood. (He's aware that there are other people trying to sell parts of the Moon but believes they are doing so "with criminal intent."[14]) Moreover, Hope says, Lunar Embassy has pulled him out of severe financial difficulty and provides a better salary than his previous jobs ever did, so he must be doing something right.

The claim, of course, is astonishing. *This guy thinks he owns the Moon? (. . . and other people buy it?)* Clearly the UN, US, and USSR found the assertion too ridiculous to merit a response. Who is Dennis M. Hope of Rio Vista, California, to say he owns the Moon? He's never even *been* to the Moon. Is the Moon something a person might own in the first place? And yet there he is on any given business day, parceling out square tracts of lunar territory (lunitory?) on a first come, first served basis.

As absurd as Hope's enterprise may seem, however, it is no less absurd—and far less destructive—than a pope's having "given" the so-called New World to Spain. And yet this particular absurdity led to Europe's total conquest of its American Canaan.

In the spring of 1492, just two months after Christopher Columbus returned from his first expedition to what he called the "other world" (*otro mondo*), Pope Alexander VI decided to hand that world over to his favorite monarchs. Ferdinand and Isabella had just succeeded in winning the previously Islamic Iberian Peninsula for Christendom and ridding it of its Muslim and Jewish citizens. So, the pope reasoned, they had proven themselves to be up to the task of winning land for Christ. Besides, now that the Reconquista was over, they probably had some time on their hands. Therefore, Alexander wrote, "we, of our own accord . . . give, grant, and assign to you and your heirs and successors . . . all islands and mainlands found and to be found . . . toward the west and south . . . from the Arctic pole . . . to the Antarctic pole."[15]

It was not an unprecedented move. Previous papal bulls had granted dominion of most of the African continent to the kingdom of Portugal on similar grounds (namely, its having conquered the Muslims of North Africa). Infamously, the continually revised *Romanus Pontifex* (1455) not only handed the African land, its resources, and trade rights over to "King Alfonso and his successors," but it also instructed the Portuguese to "reduce [African] persons to perpetual slavery."[16] So when Alexander doled out the Americas to Spain a few decades later, he imagined he was leveling the colonial playing field, giving the whole Iberian Peninsula a portion of Earth.

Just as we did with the case of Dennis Hope, who admits he's exploiting a legislative loophole, we might ask by what right these Bishops of Rome believed they owned Africa and the so-called New World, such that they had the right to give these lands away. The answer, as Alexander explains in a papal bull called *Inter caetera* (1492), is that the papacy can grant the land "out of the fullness of our apostolic power, by the authority of Almighty God conferred upon us in blessed Peter and of the vicarship of Jesus Christ, which we hold on earth."[17] The short translation: "I can give you this land because God has entrusted it to me."

The longer translation: "I can give you this land because God appeared on Earth in the person of Jesus Christ; Christ gave all earthly authority to the Apostle Peter (Matthew 16:18); Peter was the Bishop of Rome; and now *I'm* the Bishop of Rome, so I'm invested with the 'vicarship' (or representation) of Jesus Christ himself. Since Christ is the ruler of the world and I, Alexander, am the living vicar of Christ, *this land is my land*, and I'm giving it to Spain."

As Alexander explains it, the reason he is entrusting Columbus's *otro mondo* to Ferdinand and Isabella is that they have the best chance of winning that world for Christ. As the first expedition had already shown, the "very remote islands and even mainlands" of the *otro mondo* were home to "very many peoples living in peace . . . going unclothed, and not eating flesh," whose raw intelligence, moral conduct, and existing belief in a God (even if it was the "wrong" one) made them suitable for conversion.

By describing the Indigenous people of Hispaniola as peaceful, naked vegetarians, Alexander is likening them to Adam and Eve, living in a "primitive" state of innocence. The message is clear: Native Americans are essentially Europeans who are frozen in time, awaiting Christ's redemption as only Europe can dispense it.

For all Alexander's lofty proclamations about the salvation of Indian souls, however, he adds a most telling aside. The land, he tells the monarchs, happens to contain not only people but also

> gold, spices, and very many other precious things of diverse kinds and qualities. *Wherefore*, as becomes Catholic kings and princes . . . you have purposed with the favor of divine clemency to bring under your sway the said mainlands and islands with their residents and inhabitants and to bring them to the Catholic faith.[18]

The prose is overstuffed, but the "Wherefore" says it all. The land contains materials that Spain would like to possess. For this reason (*wherefore*), it is crucial to baptize the Native people. Granted, this logic is perplexing. What does the conversion of Indigenous Americans have to do with their giving the land and its riches to Spain?

The sinister connection between spiritual and economic endeavors in the New World becomes clear in a 1513 document called the Requisition (*Requerimiento*), which announced the invaders' intentions to the inhabitants of the Americas. In a ceremony that would be farcical if it hadn't had such devastating consequences, the conquerors would recite this Requisition upon landing on an unfamiliar shore. Standing at the ship's bow, or on a high point on the beach, one of the Spaniards would intone these words *in Latin*—a language very few of the sailors and absolutely none of the Natives would have known: "On the part of the King, Don Fernando, and of Doña Juana, his daughter, Queen of Castile and León, subduers of the barbarous nations, we their servants notify and make known to you, as best we can, that the Lord our God, Living and Eternal, created the Heaven and the Earth. . . ."[19]

Let me interrupt this scene to call attention again to its absurdity, which is honestly hard to overstate. When I teach this text in class, I encourage my students to imagine an alien spaceship, landing at the end of their block and opening to reveal a green humanoid who proceeds to shout half an hour of . . . instructions? Prayers? The Venusian phone book? Who knows! "You don't speak alien," I'll say, "so you honestly have no clue what this creature is going on about."

"But just for argument's sake," I'll continue, "pretend you happen to have some cosmic translation device that allows you to understand what the green fellow is shouting. You listen to the whole speech before heading back inside to your parents and siblings, who ask what on Earth is going on. What do you tell them?"

And here, the students dig out the five major points of the Requisition.

1. God created the universe, along with Adam and Eve, the ancestors of all humanity—including Indigenous Americans.
2. God has entrusted the universe to a man named St. Peter and his successors, whom we call popes.

3. The most recent pope gave this land we're all standing on to the king and queen of a place called Spain.

4. The Natives should "acknowledge the church as the ruler and superior of the whole world," which is to say, they should convert to Christianity.

5. If they give up themselves and their land, the Spaniards will protect and celebrate them, but if not, the Spaniards will wage war with them, enslave them and their children, take all of their goods, and "do [them] all the mischief and dam-age that [they] can" in the name, and with the help, of God.[20]

As you can see, things get very bad very quickly in this shore-line recitation. It may be a particularly surprising progression when you consider the pious, even generous opening: the same God who created the Spaniards created the Natives as well. In other words, we're all sisters and brothers. But notice how this declaration—that all humans are children of the same God— actually forces the Native Americans into the particular story the Europeans are telling. They are not free to say, "Well, *our* God tells us that the land belongs to us, to the buffalo, and to the rivers and mountains themselves," because the boat people are insisting that there's only one God over them all, and that he's put the Spaniards in charge.

"What would your family say," I ask my students, "if you reported that the alien had claimed your neighborhood and all sur-rounding areas for a king you'd never heard of and a jealous god he said had made you?"

"Um . . . no?" one of them responded in a particularly feisty seminar. "I think they'd probably say no. And maybe that the alien was nuts."

Or, in the words of the Cenù Indians when they came to appre-ciate what the Spanish invaders were saying, that "the pope must have been drunk" and the King of Castile "some madman" for thinking they had the right to claim a land that wasn't theirs.[21] You know, the way most people react when they hear that there's a guy in California selling off the Moon.

The Power of Nothing

The people who want to colonize outer space—by mining asteroids, terraforming Mars, and making the Moon into a galactic gas station—don't tend to talk much about their colonial forebears' miserable track record. As we will see, however, there are people both within and beyond the field of astrophysics who do acknowledge this history and who are calling for a "decolonial" or "anticolonial" approach to space exploration. One that wouldn't repeat the crimes of the past by exploiting our planetary neighbors. But these artists, activists, and scholars tend to be ridiculed by the colonial enthusiasts, who insist that settling space is nothing like settling the Earth because in space there's nothing to disturb. As Mars Society president Robert Zubrin puts it, "On Mars, we have a chance to create something new with clean hands."[22] What Zubrin doesn't acknowledge, however, is that the European settlers also sought to create something new. And they, too, believed—or at least they *said*—that their New World was empty, so their hands were also clean.

Of course, the colonizers didn't really think the land was *empty*. It clearly contained both human and animal inhabitants along with seemingly infinite quantities of what the modern world calls "resources." In fact, the resources were the reason the conquerors went there in the first place and the reason the settlers stayed. But as far as the Europeans were concerned, the New World was devoid of any formal structure or institution that might prevent its becoming their property.

The nineteenth century would come to call this assumption *terra nullius*, or "no one's land."[23] Jurists of the fifteenth century called it the Doctrine of Discovery, or simply "the Doctrine." Any land that a European nation judged to be unoccupied and unclaimed was theirs either to "purchase" (usually under coercive and misleading conditions) or to take by force. Moreover, thanks to the legacy of Canaan—intensified by the charge to "make disciples of all nations"—Christian nations could claim that it was not only their right but their *duty* to take whatever land they happened to find.

If you think back to the creation stories in Genesis, you'll remember that this God makes human beings in his image, telling them to "be fruitful and multiply" and to have "dominion" over the rest of creation. According to these stories, humans are created creators. They are told to continue the work God began by ruling and filling the Earth. As we have seen, however, the biblical narrative and its Christian interpreters prefer some of these human beings over others. God allegedly gives Canaan to Israel, Africa to Portugal, and America to Spain. So it's not just that "humanity" believes it has dominion over the "fish of the sea" and the "birds of the air" and the mammals of the Earth; it's that certain humans believe they have dominion over all other humans as well.

The claim is clearly dodgy, especially for a tradition that also asserts the shared ancestry of all humanity and the "good"-ness of all creation. Did God *really* go out of his way to destroy the Egyptians and obliterate Canaan? Did he really endorse the strip-mining of the Caribbean, the enslavement of its young male inhabitants, and the wanton slaughter of the rest?[24] Does he really want to make sure "American boots" make it back to the Moon before Russian and Chinese boots make it up there?

You can sense an ethical anxiety coursing through Deuteronomy and Joshua, which obsessively repeat the divine directive to kill everything that moves—as if to drown out any doubt or regret. You can sense it in Benjamin Franklin's insistence that the divine fire be shown on the Great Seal—to assure people that any suffering inflicted in the name of freedom is the will of God. And you can sense the anxiety in the rituals the conquerors performed when they arrived on the "virgin" shores of the "new" world, reading their Latin proclamations and planting their crosses and flags.[25] But what was the point of these strange little ceremonies? Why shout words that no one understands, and what good does it do to put a stick in the ground?

The comparative religionist Mircea Eliade interprets these recitations and plantings as symbolic re-creations of the world.[26] Through ritual practice, he explains, an invading nation turns the "chaos" of the foreign zone into an ordered world, or "cosmos"— just as their God spoke and shaped the world into being at the dawn of time. The conquest of territory, in other words, is an imitation of divinity, with the conquerors claiming the power of God by enact-

Figure 3.3 Henry Sandham, illustration (1905) of Vasco Núñez de Balboa hoisting the Spanish flag in Central America. From Edith A. Browne, *Panama* (London: A. and C. Black, 1923), 48.

Figure 3.4 Giovanni Battista Carlone, fresco of Christopher Columbus planting cross in the New World. Palazzo Ducale, Genoa, Italy.

Reuters

ing that power. And the two major mechanisms of this ritual imitation are the retelling of the creation story and the installation of a vertical structure in the ground.

Why the creation story? Because it establishes the conquerors as God's people and identifies the conquest as his will. Because, as we saw with the Requisition, this creation story makes resistance seem futile by drawing the whole world into the same cosmic drama. Because the biblical narrative seems to inscribe the conquest into the order of the world itself, informing the people of the Americas that their death or defeat is inevitable.

Why the flag or the cross in the ground? To connect the land in question to the heavens above. To reinforce the conquering nation's horizontal expansion with the vertical authority of God. (Think the Washington Monument, the Great Buddha of Thailand, Christ the Redeemer in Rio de Janeiro, or the reconstructed One World Trade

Center: if you want to seem powerful, put a tall thing in the ground.) Through these gestures, the conquerors began the work of creating a world—like the God they said had sent them on their mission— out of a "wilderness" they insisted was effectively, legally, *nothing*.[27]

Destiny Made Manifest

In the winter of 1763, a gang of white vigilante frontiersmen murdered twenty Susquehannock men, women, and children in Conestoga, Pennsylvania. Calling the Susquehannock "red Canaanites," the vigilantes insisted they were executing a divine command to cleanse the contested area of its Native inhabitants. Upon hearing the news, Benjamin Franklin was horrified. "It seems that these People think they have a better Justification [than the law]," he fumed; "nothing less than the *Word of God*. With the Scriptures in their Hands and Mouths, they ... justify their Wicked- ness, by the Command given to *Joshua* to destroy the Heathen."[28]

Thomas Paine was also enraged by the common appeal to Joshua to justify the slaughter of Indigenous Americans, insisting that the Bible could not be read literally by anyone with a shred of moral sense.[29] Thomas Jefferson expressed the same moral doubt, liter- ally cutting from a few Bibles those passages he believed to be ratio- nally and ethically sound, pasting them into his own "Life and Mor- als of Jesus of Nazareth," and leaving the rest behind. The "rest" included the entirety of the Hebrew Bible, whose God he believed to be vengeful and cruel.[30]

Despite these protests, however, the image of America as God's New Israel wouldn't budge. In fact, the very men objecting in this instance to the Canaanite imagery also contributed to it, what with Franklin's Red Sea and Jefferson's Sinai desert.[31] Even the notori- ously anti-Christian Paine can be said to have advanced America's self-understanding as a "chosen" nation by secularizing the idea. For Paine, the new nation did, indeed, have a special mission on Earth; it's just that there was no active, personal God who'd dispatched it. Despite his infamous "deism" (the idea that God created the world but does not intervene in its affairs), Paine is often cited as an early architect of "Manifest Destiny," the nineteenth-century rendition of America's ever-renewed Canaan complex.

*   *   *

The last time I mentioned Manifest Destiny, Donald Trump was laying out his vision for America's future in outer space. Having "always been a frontier nation," Trump argued, America is again being called to settle a wild new frontier and embrace its "manifest destiny in the stars."[32] The idea is at least as old as the Mayflower, but the term itself was coined in 1845 as the nation debated the annexation of Texas, Oregon, and California. In the words of the editor of New York's *Democratic Review*, annexing Texas would begin "the fulfillment of our *manifest destiny* to overspread the continent allotted by Providence for the free development of our yearly multiplying millions."[33]

The editorial packs a lot into this little clause. First, it declares not just the existing states but the whole *continent* to be the "allotment" of "Providence," or the will of God. Second, it ties this land bequest to the "multiplying millions" the white folks are producing (in line with the divine imperative to "go forth, increase, and multiply" and the divine promise to make the chosen people "as numerous as the stars"). Finally, it calls America's destiny "manifest," which is to say, visible. Obvious, even. And it's here that the author claims American superiority over the entire theological tradition that produced God's New Jerusalem.

Unlike God's covenant with Abraham, which the biblical text leaves unexplained and arguably unfulfilled, and unlike God's salvation or damnation of any particular soul, which also remains inscrutable, God's election of America is said to be *manifest*. In other words, it is clearly the case. God has chosen America not out of the infinite mystery of his will but "because of [America's] superior form of government, its geographical location, and its beneficence,"[34] and God has already rewarded his chosen nation with the promises of Abraham: a land rich in resources and more descendants than anyone can count. No one could doubt that God loves the United States.

Again, when the ideology hits this hard, one wonders what it's covering up. And in this case, as in every previous stage of US history, what Manifest Destiny obscures is the claim that Indigenous Americans (and in the case of westward expansion, the nation-states of England, France, and Mexico) had already got to the land white Americans were demanding. As in previous generations, the ideologues justified their right to the West by insisting there was "nothing there" worth respecting. Just, in the words of Andrew Jackson, "a country covered with forests and ranged by a few thousand savages." Or, as William Henry Harrison put it, "a state of nature, the haunt of a few wretched savages."[35]

Considering the raw, undeveloped, and effectively unpopulated nature of the westward expanse, the original "manifest destiny" editorial assured its audience that the United States could subsume the continent's "untrodden space, with the truths of God in our minds, beneficent objects in our hearts, *and with a clear conscience unsullied by the past.*"[36] This is the same sort of "clear conscience" that space colonizers now assure us we can have for real this time, because after all, there's nothing there to disturb. But one might ask the spaceniks the same question we'd ask the frontiersmen and conquistadores of yore: if there's nothing there, then why do you want the land in the first place?

The answer, of course, is and was and always will be "resources." Gold, spices, fur, ore, helium-3, hydrogen, platinum, animal flesh, human labor—whatever might open a new economy for the benefit of the extractors. As far as the American settlers were concerned, Indigenous nations "had let their resources go to waste,"[37] neglecting or even refusing to "own" and "improve" the land as God directed Adam when he told him to "till the earth and keep it" (Genesis 2:15). And so as the West was "won," it was lost to its traditional caretakers, who were driven from their ancestral lands onto reservations. Even these areas were reduced and relocated when white settlers found them to be more valuable than they'd initially calculated. Cut off from their land, unable to move freely, corralled with rival nations, and often forced to convert, wear European dress, abandon their languages, and attend English schools, Indigenous Americans faced attempted extinction at the hands of the people of God.

You shall annihilate them—the Hittites and the Amorites, the Canaanites and the Perizzites, the Hivites and the Jebusites—just as the Lord your God has commanded, so that they may not teach you to do all the abhorrent things that they do for their gods (Deuteronomy 20:17–18).

From my perspective, the most unbearable testimony to the power of this text as it shores up American chosenness is the 1854 oration of Duwamish Chief Seattle. Upon ceding his people's land to Governor Isaac Stephens of Washington Territory, Chief Seattle told the white men, "Your God is not our God. Your God loves your people and hates mine." The reasons might be hidden, but the effects were, well, manifest: "Your God makes your people strong every day. Soon they will fill the land. Our people are ebbing away like a rapidly receding tide that will never return."[38] What's particularly awful about this oration is Chief Seattle's seeming acceptance of the vast mythic pretense that's decimated the continent. God must be on your side, he reasons; otherwise, how could you possibly be so strong?

Being an adept rhetorician, however, Chief Seattle seems to know that the conquest of the Americas does not actually testify to the power of God. Rather, it testifies to the power of God-language. Specifically, the American conquest affirms the power of invoking divine support for the seizure and stripping of land. What I'm trying to say is that it doesn't matter whose side God was *actually* on, or even what the conquerors and founding fathers *actually* believed. (How on Earth would we know?) What matters is the real-world work their mythic frameworks have done regardless of their falsity or truth. "God's New Israel" enslaved, displaced, and murdered its way across the entire continent while announcing its unparalleled "beneficence," thanks to the world-ending, world-making power of myth.

Then, with the dawning of the space age, God's New Israel set its sights on the heavens themselves.

CHAPTER FOUR

The Final Frontier

There is something more important than any ultimate weapon. That is the ultimate position—the position of total control over Earth that lies somewhere out in space.
LYNDON B. JOHNSON

Screwing Up the Moon

In the late 1960s, NASA leased a large tract of Navajo land to test the module that would deliver Neil Armstrong and Buzz Aldrin to the lunar surface. The resulting interaction has been retold in so many different ways that it's obtained legendary status, but the gist is that a Navajo singer asked a NASA official to deliver a recorded message to the people of the Moon. When the official asked the chief to translate the singer's message, the chief stalled a bit before saying, "He's telling the moon people to watch you guys carefully, because you might screw things up on the moon the way you have on Earth."[1]

If the joke seems funny—if it even comes across as a joke—it is perhaps because the Apollo missions have shown that there *are* no "moon people." No one to infect with earthly viruses, enslave in lunar mines, or displace to the craters the Sun doesn't reach. But the apparent lack of Indigenous people on the Moon does not erase the immeasurable damage done to Indigenous terrestrials in the name of discovery, freedom, and destiny. And in this context, the Navajo singer's message to "the Indians of outer space" can be heard as a mournful satire of the strategies of European conquest and US expansionism, which had already found an eerie reprise at the dawn of the space age. The moment America set its sights on space, the old imperial game was on again.

The last chapter traced a persistent analogy in early colonial journals, sermons, and political speeches between the land of the Americas and the land of Canaan. While the Old World was likened to Egypt, the New World was coded as Jerusalem, with European Americans proclaiming themselves to be God's New Israel. Under a perceived divine mandate, these settlers asserted their right to take, mine, and recreate the entirety of North America for the glory of God.

In 1890, the US Census Bureau declared this chosen nation's frontier to be "closed," with most of its arable land parceled out to non-Indigenous people through the Homestead Act. Just over half a century later, a tangle of politicians and rocket scientists proclaimed the frontier "open" again as the US raced the Soviet Union to the Moon. And sure enough, this renewal of American frontierism was guided by a renewed sacred analogy. Just as the sixteenth century's New World was said to be a new Promised Land, the twentieth century's "outer space" was said to be a new New World: a place of adventure, danger, political freedom, economic opportunity, and endless promise.

For what it's worth, the Cold War did not invent this particular analogy. Seventeenth-century naturalists were already comparing the Moon and other planets to the Americas in the wake of Galileo's telescopic discoveries. Against anyone who thought it ridiculous to speculate about the inhabitants of Venus or to imagine traveling to Mars, these early popular science writers brandished the outlandishly heroic figure of Columbus, who sailed a sea no one thought it possible to sail, found land where no one expected land to be found, and discovered people no European would have dared to imagine existed.[2] Sure, it seemed absurd that humans might one day make it to the Moon, or that lunarians might make it to Earth. But had Native Americans thought it any less absurd when Europeans suddenly landed on their shores, claiming the land (in Latin) for a God they'd never heard of?

Beginning in the mid-1950s, this analogy between the seas and the spaceways came roaring back. As the Cold War escalated, an outpouring of editorials, political speeches, and educational films likened the impending American journey through space to the pilgrims' journey across the Atlantic and their descendants' trek across the West. United States astronauts would be nothing less than new pioneers, embodying that distinctly American desire— some even called it a *need*—to move ahead, push forward, explore, and *expand*.[3] And although the references to God became a bit subtler over the course of the twentieth century, this astrofrontierism was grounded in earthly frontierism, which itself was grounded in biblical land claims. In short, by comparing outer space to the New World, early space enthusiasts were projecting America's self-perceived divine mandate out to the heavens. Just as God had called the Israelites into Canaan, the Europeans into Connecticut, and the Homesteaders out to Colorado, so was he now—in the face of its Soviet rival—calling America to the Moon.

Luna Americana

The creation story of the American space program usually begins with the trauma of Sputnik. In October of 1957, the USSR launched Earth's first artificial satellite and then repeated the trick a month later with a sacrificial dog named Laika on board. Within days of the first launch, the Eisenhower administration began assembling a committee that, less than a year later, would establish the National Aeronautics and Space Administration.

Surrounded as we now are by hundreds of thousands of functioning and defunct pieces of orbital machinery—including most notably the International Space Station—it can be difficult to recreate the amazement, fear, disgust, and raw shock that the Sputnik launch inspired.[4] For billions of years, the night sky was flecked with the same ethereal planets, moons, asteroids, and stars—and then suddenly it also contained *a thing that human beings had made*: a 184-pound metal sphere with four skinny antennae pulsing out signals to Earth. In Russian, *sputnik* means "friend," "traveling companion," even "spouse." Or, in astrospeak, "satellite."

"What do you think of our two new moons?" the German philosopher Hannah Arendt wrote to her friend Karl Jaspers after the second Sputnik launch. "And what would the moon [itself] likely think? If I were the moon, I would take offense."[5]

The US certainly took offense. Not on behalf of the Moon, of course (I'll repeat that, unlike many Western *people*, Western *institutions* don't tend to personify land), but in the name of its own security and global supremacy. Space had suddenly become a new frontier, and America was already running behind. As Lyndon Johnson, then Senate majority leader, fumed, "The Roman Empire controlled the world because it could build roads. Later—when men moved to the sea—the British empire was dominant because it had ships. In our age we were powerful because we had airplanes. Now the Communists have established a foothold in outer space."[6] So the Capitalists were going to have to catch up.

It was arguably President John F. Kennedy who formalized this astronautic "race" by imposing a deadline. Under the considerable influence of Vice President Johnson, Kennedy announced in the spring of 1961 that the US intended to send a crewed mission to the surface of the Moon and back "before this decade is out."[7] As Kennedy had argued on the campaign trail, everything was at stake "in this vital race" to outer space, from political dominance to military positioning to economic sovereignty to the freedom of religion, which the Communists would gladly abolish if the US let them win.

"If the Soviets control space," Kennedy insisted, "they can control earth, as in past centuries the nation that controlled the seas dominated the continents."[8] There's that analogy again between the frontiers of Earth and space, between the domination of the continents and the control of the cosmos. As the US faced its atheist-Communist adversary, the nation's old religious imperialism woke up, dusted itself off, and took a sharp, vertical turn toward the Moon.

Clearly, then, the US space program was born in a crisis over military, political, economic, and even religious dominion. The US went to space to secure what Johnson explicitly called "the position of total control over Earth."[9] At the same time, the very speeches and legislation that announce this intention also surround it with more humane, even noble companions—like exploration, scientific discovery, human advancement, and world peace. This is the "doublespeak" that converts, displaces, and even enslaves people "for their own good," and it goes cosmic in the wake of Sputnik, when President Eisenhower declares that the US must develop a *national* space program "for the benefit of all mankind."[10] In the duplicitous logic of the space race, American military dominion does not merely exist alongside the more benevolent pursuits of science, adventure, freedom, internationalism, and peace: it *enables* them. American dominance for all mankind.

The clearest example of this persistent contradiction can be found in the speech that Senator Kennedy gave during the 1960 presidential race. Having just asserted that "If the Soviets control space they can control earth," Kennedy nevertheless insisted, "This does not mean that the United States desires more rights in space than any other nation. But we cannot run second in this vital race. To insure peace and freedom, we must be first."[11]

This is one of those passages I have to read a few times to make sure I'm not missing something. On the one hand, Kennedy is saying that the US just wants the rights of any other nation in space. On the other hand, he's saying that the US has to win. Is he suffering some sort of cognitive lapse? How can the US both share, take turns, be a good neighbor, *and* insist that it has to go first? Even a kindergartener could see that these rules aren't fair.

The solution to this riddle lies in that Cold War belief that "America" incarnates the freedom that "Communism" destroys. This belief is a direct descendant of Manifest Destiny, which set the US apart as an example of peace and freedom to the rest of the world and which itself inherits the old biblical ideas of a chosen nation and a Promised Land. When Kennedy says that "to insure peace and freedom, we must be first," he is plugging into that centuries-old idea that America is "God's New Israel," destined to lead the other nations in the ways of justice and peace.

So this is the way to hold together Kennedy's visions of global equality on the one hand and American dominance on the other: if the US *is* peace and freedom, then peace and freedom lose if the US loses and win when the US wins. They win so completely, in fact, that even the Russians will come around, subjecting themselves to the benevolence of American leadership. As Kennedy told American taxpayers when he asked them to fund the Apollo missions, "we are anxious to live in harmony with the Russian people. . . . We seek no conquests, no satellites, no riches. . . . We seek only the day when 'nation shall not lift up sword against nation, neither shall they learn war anymore.'"[12]

The ensuing decades of surveillance, suspicion, and direct sabotage would come to contradict most of Kennedy's assurances here, but the part of the speech I'd like to focus on is his final quotation of the Hebrew Bible's book of Isaiah. In it, the prophet is imagining a peaceful future for Jerusalem: a Promised Land living up to its promise. But as the book's first chapter details, this peace will be preceded by God's purging the city of every unfaithful inhabitant, scouring its streets of all miscreants, and judging among the nations. Once the city is purified, it will become the seat of God's kingdom, where the people "shall beat their swords into plowshares" and study war no more. But until then, it's going to be a nasty battle between the righteous and the unrighteous.

So even as President Kennedy envisions "harmony" with America's worst enemy, he envisions it on America's own, biblical terms. "We seek peace," the Catholic liberal says in the face of the godless Communists; specifically, we seek the peace of the God of Abraham, Isaac, and Jacob—who incidentally favors this nation over everyone else.

In Kennedy's citation of Isaiah, we can once again witness religion being dragged in to cover up otherwise questionable intentions. Just as Spain insisted it was conquering the Americas to save souls, just as the pilgrims framed the slaughter of Native Americans as a holy war, just as plantation owners claimed they were Christianizing the people they enslaved, and just as expansionists declared that "Providence" had given the entire continent to white settlers, American politicians and scientists alike have justified the space program since its inception by appealing to the will of God and the sacred destiny of humanity. After all, military dominance and national glory only go so far with the American public. If you really want to drum up excitement for a war, an annexation, or a trip to the Moon, you've got to sell it as a sacred duty.

A good deal of the space program's religious force comes across in the very language of "the frontier," which as we have seen, imagines certain people as divinely ordained to occupy faraway lands. Thanks to the founding myth of Canaan, revived in the Doctrine of Discovery, Manifest Destiny, and the moon shot alike, even those seemingly secular appeals to "the spirit of discovery" or "man's need to conquer new terrain" rely on a fundamentally religious logic of chosenness. And because it's so fundamental to the American endeavor, religion bubbles up in space-talk all the time—the way it does in that Pence speech, or that Trump speech, or that Kennedy speech . . . or in the Apollo missions.

The Apollo missions? Yes. I'm going to argue that, as thrilling as I still believe them to be, the Apollo missions replay the Christian imperialism that founded and expanded the US. In fact, I'm going to argue that these missions function "religiously" even beyond their being named after a god and even beyond their being called "missions." But to arrive at this conclusion in its full force, we're going to have to take a detour through Disneyland.

Finding Tomorrowland

The year was 1954. Walt Disney had clear-cut 160 acres of orange and walnut trees in Anaheim, California, to construct a massive theme park. He'd planned the park around four anchoring "lands": Fantasyland would be centered on Sleeping Beauty's castle; Adventureland would replicate "the remote jungles of Asia and Africa"; Frontierland would put the visitor in a Davy Crockett cap and lead him via covered wagons, pack mules, and a mine train through the wilderness; and Tomorrowland would look like . . . actually, Disney had no idea. What would the future look like? What *could* it look like? Disney was apparently stuck. What could he put in Tomorrowland?

Disney consulted his animators, one of whom suggested he look to a recent issue of *Collier's* magazine about the possibility of space travel. On its cover, a rocket plane traverses the earthly horizon, leaving a bright red blaze behind it. In its pages, the nation's top physicists explain the current state of rocket science and insist that if the US doesn't embark on a formal program to attain "space superiority," the USSR will beat them to it.[13] This eventually became the argument Senate Majority Leader Johnson used after the launch of Sputnik to spur the creation of NASA and that President Kennedy used to insist that America be the first to reach the Moon. But out in Anaheim, the space race was doing less politicizing than imagineering, building the popular infrastructure of America's God-given future in space.

The chief contributor to this *Collier's* issue was Wernher von Braun, the former Nazi rocket engineer who was granted amnesty in 1945 in exchange for his scientific service to the US military. Von Braun's lead article was called "Crossing the Last Frontier," and it inspired everyone from Dwight Eisenhower to *Star Trek*'s Gene Roddenberry to extend the metaphor (you don't have to be a Trekkie to shout back, "*Space.* The final frontier"). And even in von Braun's original formulation, the image of "the last frontier" was calibrated to hit with its fullest emotional, political, geographical, technological, and yes, religious force.

In the short years following his immigration by means of the CIA's then-secret Operation Paper Clip, Wernher von Braun had become a fervent believer in the *Pax Americana*. As he integrated his newfound US patriotism with his astrophysical aspirations, he became convinced it was America's duty to extend Western civilization not just across the globe but into outer space. And just as generations of European Americans had done, von Braun justified his imperialism by calling it a vehicle of freedom, democracy, and eternal salvation.

If that last bit comes as a surprise, it's because very few people tend to talk about von Braun's postwar conversion to evangelical Christianity. During his de-Nazification on American soil, the rocket scientist became "born again" in nearly every sense of the term, ultimately deciding it was incumbent on the US to spread the Gospel not just around the world but to "the heavens themselves."[14]

In other words, von Braun explicitly framed US space travel as a vertical extension of Manifest Destiny. Just as God had allegedly endorsed and even demanded the westward expansion in the nineteenth century, God was calling European-descended Americans in the twentieth to conquer a new frontier—a bigger frontier, even an infinite frontier—in outer space.

How did this former German citizen know so much about American Manifest Destiny? Unfortunately, this particular road passes through Adolf Hitler, who defended his plan to swallow Eastern Europe by comparing it to the US westward expansion. Hitler's notion of *Lebensraum*, or expanded "living space" for the burgeoning German people, was an updated German version of the Manifest Destiny that had destroyed and displaced Indigenous America. In fact, as he invaded Russia, Hitler declared, "There's only one duty: to Germanize this country by the immigration of Germans and to look upon the Natives as Redskins."[15] In von Braun, then, a Christian American political doctrine was filtered through its Nazi German revival to produce a born-again American astrophysical dream of colonizing the cosmos, a dream powered by the V-2 rockets built by enslaved concentration camp workers to destroy mid-century London, Paris, and Antwerp. And it was these rockets that finally solved Walt Disney's Tomorrowland problem.

The minute he read the *Collier's* issue, Disney summoned von

Braun to Anaheim. The scientist's rockets would need some aesthetic work—Disney couldn't imagine public pilgrimages to anything so ugly—but the infamously shape-shifting von Braun was happy to take Disney's notes and reshape his missiles into the more streamlined, Jetsonian projectiles that would rise from the center of Tomorrowland. Disney would furthermore position Tomorrowland directly opposite Frontierland, presenting America's future in space as an extension of its pioneering past.[16]

Once Tomorrowland had found its tomorrow, von Braun served as professor-expert in two major Disney films about the impending human "conquest of space." Thanks to the small number of television channels and the staggering novelty of the topic, it is estimated that half of all Americans ended up watching these films, which presented outer space as their next home. Thanks to the vision of von Braun, space would be the final step in a holy saga that had led "man" from his primitive caves across fertile plains to ancient empires, medieval Europe, the shores of America, and eventually the wilds of California—which had now become the de-wilded site of Tomorrowland. And in just ten years, von Braun's aesthetically upgraded rockets would model the Saturn V that brought the Apollo crews to the Moon.

As religion scholar Catherine Newell argues, it was therefore Disney and von Braun who drummed up the popular support JFK needed to escalate the space race.[17] By plugging into the sacred myth of the American frontier, Disney's park and films presented outer space not as the site of military rivalry, endless surveillance, nuclear escalation, or even improved telecommunications but as the extraordinary destiny of ordinary people. Folks like you and me, who will one day live with our movies and theme parks on Mars.

This Space Is Our Space

On Christmas Eve, 1968, one billion earthlings tuned in to see live footage of the Apollo 8 mission around the Moon. The orbit was exquisitely timed; as NASA's public affairs officer Julian Scheer figured, the night before Christmas would find most Americans at home with their families. Knowing that more people would be watching this broadcast than any previous event in history, the crew showed the viewers around a little (here's this crater, here are those cracks, there's the Sea of Tranquility), but honestly, all anyone could see was a gray quasi trapezoid that could just as well have been someone's kitchen counter as the surface of the Moon. Suddenly, without any change in the image, the mood of the broadcast becomes measured and formal as astronaut Bill Anders announces the lunar sunrise. Presumably, he and his colleagues have just seen the coming of the light.

"And for all the people back on earth," says Anders, "the crew of Apollo 8 has a message that we would like to send to you":

> In the beginning, God created the heaven and the earth. And the earth was without form, and void, and darkness was upon the face of the deep. And the spirit of God moved upon the face of the waters. And God said, "Let there be light," and there was light.[18]

Anders, Frank Borman, and Jim Lovell take turns reading the whole first book of Genesis. You know, the story about the six-day creation, the goodness of every part of it, and the making of humans in the image of God. From their spaceship orbiting the Moon.

No matter how many times I think it through, this ritual makes me pause. Why did the crew of Apollo 8 read *Genesis*? Why did they read anything at all? The most straightforward but least satisfying answer is that as Julian Scheer prepared the astronauts for the mission, he told them to "say something appropriate," since it was such a momentous occasion.[19] After asking all manner of colleagues, friends, and their spouses, the crew decided on Genesis. But what was it that made them judge Genesis to be the appropriate text? Why not the national anthem, or Mancini's "Moon River"? Why not the Navajo message to the people of the Moon, warning them to watch the white guys closely?

Here I can't help but generate another split screen, with Apollo 8 orbiting the Moon on one side and Yuri Gagarin orbiting Earth on the other. The first human to travel to space, Gagarin is said to have exclaimed on his 1961 journey, "I see no God up here!" Whether or not he actually did, the Russian press popularized the story of this antirevelation, gleefully enumerating all the saints and angels Gagarin hadn't seen in his trip around their dwelling place. More importantly, the press attributed the Soviets' clear supremacy in space to their rugged atheism. As one atheist journal explained, the reason the West lagged so far behind was that it was stuck in "dark superstitions" while the USSR was off "storming the heavens."[20] Religion was holding the Western world back from the revelations of modern science.

In this light, the Apollo 8 reading of Genesis was "appropriate" because it announced America's "Judeo-Christian" victory over its Soviet-atheist rivals. If the Russians couldn't see God in the heavens, then the Americans would just bring him on up there.

But why God? Why not Adam Smith or Alexander Hamilton or Henry Ford or a gold-embossed investment portfolio? Surely atheism wasn't the only feature of Communism the Americans found objectionable. Why target the religious difference rather than the political or economic difference? What's God got to do with it?

At this point, it's important to remember that in the Americas in particular, religious behavior and religious claims have traditionally underwritten political and economic pursuits. Such behaviors and claims include the Spaniards' reading of the Requisition, Columbus's erection of crosses and banners, Franklin's Exodus scene on the Great Seal, and any State of the Union address that asks God to "continue to bless these United States." Just like these other rituals, Apollo 8's reading of Genesis laid claim to a divine mandate. By reminding the world that their God had created the heavens and the Earth and put "humans" in charge, the crew of Apollo 8 was asserting America's God-given dominion over creation itself. This space, they were saying, is our space.

What clearer confirmation could there be than Apollo 11, with its ceremonial planting of the American flag? Directed to do so by Congress, but knowing there was no air on the Moon in which the flag could wave, engineers at the Johnson Space Center in Houston wove a metal rod into the fabric of the flag to ensure the whole Earth could see the stars and stripes extended from the pole. They apparently used the wrong heat-resistant coating, so when Armstrong and Aldrin went to extend the rod, it was bent. "Even better," thought the engineers and senators back home; the flag appeared to be waving in the "wind," announcing America's triumph to anyone with access to a television. Meanwhile, Michael Collins circled the Moon in a command module named after Christopher Columbus.

At every turn, then, Apollo was announcing its extension of earthly imperialism. Just as the Europeans had claimed the Americas by reading a sacred text and planting a cross or flag, the Apollo missions claimed the Moon with a recitation of Genesis and a highly choreographed raising of the star-spangled banner. (Six of them, in fact: each of the US lunar landings culminated in a ceremonial planting of another flag.) Of course, the Apollo missions couldn't technically claim the Moon. As the next chapter will explain, the US was technically subject to an international treaty declaring that no nation could appropriate "the Moon [or] other celestial bodies."[21] As Congress sought to justify its directive in the face of this treaty, it therefore assured the international community that the gesture would be purely symbolic. A loving testimony to its recently slain president, who had challenged NASA to get to the Moon and back by the end of the decade. So no, the Moon doesn't and can't—at least for the time being—belong to any nation. But symbolically, it belongs to the US, because the US got there first.

The US got there first, as JFK had insisted. But as JFK had also insisted, the US got there first "for all mankind." It says so on the plaque Apollo 11 left on the Moon, signed by Neil Armstrong, Michael Collins, Buzz Aldrin, and Richard Nixon and announcing, in all caps,

Figure 4.1 Neil Armstrong and Buzz Aldrin planting the American flag on the Moon.
NASA

HERE MEN FROM THE PLANET EARTH

FIRST SET FOOT UPON THE MOON

JULY 1969, A. D.

WE CAME IN PEACE FOR ALL MANKIND.

This strange silver rectangle, bent across the body of a defunct lunar lander, embodies the tension we've detected throughout this chapter between American nationalism and benevolent universalism. By leaving the plaque (not to mention the flag), Apollo 11 is making it clear that the US made it to the Moon before anyone else. But the *text* of the plaque asserts a broad humanitarianism—a "peace" and fraternity that the very *fact* of the plaque belies.

A rough translation would be something like, "look at this awesome thing that America did all by itself for the sake of everyone else!" No citizen of any other nation could possibly believe the altruism, what with that American flag "waving" in the nonwind. I even find it hard to believe that Americans *themselves* could believe it. So the message seems to be addressing no one at all.

Unless, of course, the plaque is addressing someone other than "mankind"? Some hypothetical extraterrestrial, wondering what the earthlings think they're up to on the Moon? That, at least would make the plaque's declaration intelligible: maybe the US is assuring any *aliens* who might be out there that they've come in peace, for all mankind. But exactly which aliens do they think will be able to read the message? How seriously would a Martian take those earnest astronautic and presidential signatures? Announcing the munificent intentions *and* unilateral authority of a bunch of people from a faraway world, the lunar plaque looks strikingly like the Requisition. Just like the Requisition, the plaque claims an unclaimable land in an indecipherable language on behalf of rulers no one in the new world would recognize.

And in that case, it could be that the Apollo missions ended up delivering the Navajo message to the people of the Moon after all. Because these rituals seemed to announce that the US would indeed screw up the heavens with the same flags, ceremonies, and humanitarian justifications that screwed up the Earth.

Missions of Mass Distraction

When Jeff Bezos returned to Earth in his cowboy hat during the summer of 2021 on the heels of Richard Branson, he faced a torrent of criticism from nearly every news outlet and social media platform. Wildfires were ravaging the Pacific Northwest, floods were devastating India and Germany, the Delta variant was killing off systematically misinformed Americans, and this Wild West billionaire *left the planet*? In a fuel-burning rocket-phallus that scorched the ground it departed from and polluted the air that carried it?

Predictably, Bezos responded by saying his space company would eventually help address all these crises. After all, Blue Origin was paving the way for an off-Earth world. "We have to build a road to space," he explained again, "so that our kids and their kids can build a future."

"*Throughout your generations forever.*" It's the keystone of God's promise to Abraham, Moses's charge to Joshua, Pope Alexander VI's bequest to Ferdinand and Isabella, the founding fathers' rebellion, the Homestead bequests, the moon shot, and now the corporate American clamor for outer space: I'm bringing you to a land where your children and your children's children can finally be free. This is a fantastic rhetorical strategy, because who could possibly object? Who doesn't like freedom, or children, or the future? But as we keep seeing, when the ideology hits this hard, it's probably concealing something sinister. Like genocide, enslavement, dislocation, arms escalation, or in the case of the astropreneurs, exploited labor, climate abuse, tax evasion, and just good old-fashioned greed.

Thanks to the rhetorical force of phrases like "the future of humanity," "a multiplanetary species," "the light of consciousness," and "the spirit of exploration," the space barons are trying to convince us that the prodigious damage they're doing on Earth is instrumental to their messianic missions. That the tens of thousands of delivery trucks polluting the atmosphere for Amazon are funding the eventual Earth-saving relocation of all heavy industry to some asteroid, somewhere. That the reefs Musk obliterates in his failed rocket launches are a necessary sacrifice for the wicked snorkeling adventures we'll have on a terraformed Mars. But to return for a moment to the recurring phantom of religion in this space race, the only way to invest in the vision of these techno-prophets is to have a heavy dose of *faith* in stuff no one's ever seen. Like rotating space pods, for example, or Martian potato farms—or a society of true equals, toughing it out under corporate control without reliable access to oxygen.

As we've already seen, Bezos and Musk are hardly the first people to imagine such utopias. Bezos got his cylindrical idea from his Princeton professor Gerard O'Neill, author of a 1976 popular science book about orbital space colonies. Titled *The High Frontier* (there's that omnipresent metaphor), O'Neill's book opens with a fictional letter from one leisured, middle-class, heterosexual couple to another extolling the shopping, restaurants, and entertainment on a space cylinder calibrated to have "a Hawaiian climate."[22] Musk, in the meantime, is following the ideological lead of Mars

Society CEO Robert Zubrin, whose epilogue on "the Martian frontier" warns that Western society will disintegrate unless it finds a "New World" to colonize.[23] And of course, O'Neill and Zubrin are building on the decades of frontierism that von Braun, Disney, Johnson, and JFK revived at the dawn of the space age. So no, Bezos and Musk didn't design these extraterrestrial dreams. They're just the first earthlings rich enough to try to realize them.

Well, not exactly the first. As space reporter Christian Davenport has shown, the late twentieth century saw numerous entrepreneurs attempt to privatize the space industry only to fail in the face of a contractual stranglehold between NASA, Boeing, and Lockheed Martin.[24] What sets Bezos and Musk apart, along with Branson the slightly lesser baron, is not just their relentlessness but also their timing.

As we saw in chapter 1, it was Barack Obama's 2011 NASA budget that enabled the explosion of NewSpace by canceling the space shuttle program and passing the baton to the private sector. The stated hope was that this policy would allow the space industry to grow independently of future government subsidies and relieve taxpayers of most of the burden of space exploration. As it turns out, however, taxpayers are still funding the multibillion-dollar government contracts for which NewSpace corporations are so fiercely competing. And it's not clear that anyone's saving money apart from their CEOs, who have received—in addition to the contracts—grants, tax breaks, and subsidized loans. Leftist congressfolk are so frustrated they've started calling these handouts "welfare for billionaires."

At the time of my writing this book, for example, NASA had just awarded $2.89 billion to SpaceX to return the US to the Moon through the Artemis mission. Incensed, Jeff Bezos lobbied Congress to grant Blue Origin a second contract so that, as he explained in an open letter to NASA administrator Bill Nelson, "two competing lunar landers" might keep prices down and quality up.[25] The price of keeping prices down is currently $10.032 billion, which Bezos supporter Senator Maria Cantwell of Washington proposed adding to the Senate's bipartisan "Endless Frontier Act" (you can't make this stuff up) so that NASA would have enough money to give to SpaceX *and* Blue Origin. Meanwhile, the nation's pandemic-weary citizens received a meagre $600 after a year of congressional stalling. Meanwhile, Musk idiotically brag-tweeted that Bezos "can't get it up."[26]

In the torrent of news coverage following Jeff Bezos's summertime spaceflight—slightly longer than Richard Branson's and on an even more anatomically suggestive vessel—most missives called our attention to the emerging industry of space tourism. How dare these obscenely wealthy men respond to our global catastrophes by creating cosmic joyrides for other obscenely wealthy men? Friends who hadn't been paying much attention suddenly texted me about Virgin Galactic's lounge-lizard spaceplane, Blue Origin's tourist-dummy Mannequin Skywalker, one proposed pill-shaped space hotel where $9.5 million gets you twelve nights but no food, and another that's going to be wheel shaped and named after Wernher von Braun. Incensed microbloggers recirculated that horrible interview where Bezos said he couldn't think of anything to do with his "Amazon winnings" but travel through space.[27]

Yes, the prospect of orbital tourism is odious—even obscene. But it's just the tip of an even more threatening spaceberg. In between the astropreneurs' short-term, near-Earth dreams of zero-g joyrides and their long-term, utopian promises of human salvation lies a massive tangle of technological, economic, and military pursuits, arguably the real stuff of this public-private partnership. A sprawling, political space beast whose scope and intentions are reliably masked by farcical visions of space getaways on the one hand and tragic promises of salvation on the other.

Space Pioneers

Just a few months after the 1957 Sputnik launches, President Eisenhower's Science Advisory Committee published an "Introduction to Outer Space" for the American people.[28] Seeking to justify the

considerable cost of developing a space program, this short document lists "four factors" that lend outer space its "importance,
urgency, and inevitability."

1. Exploration (or, as the committee phrases it, "the com-
    pelling urge of man to explore and to discover . . . to go
    where no one has gone before"—even more fodder for
    Roddenberry)
2. Defense
3. "National prestige"
4. Science

As the authors demonstrate, these four chief endeavors are all
bound up with one another. The pursuit of science relies on the
explorers' raw curiosity; curiosity is often driven by a desire for
prestige; and prestige is augmented by a nation's exploratory, scientific, and military accomplishments. "In fact," the scientist-authors
admit, "the quest for ultra long-range rockets" was crucial to the
development of modern space science in the first place. Ultimately,
then, the whole infrastructure of the space program rests on military interests and military spending.

Three decades later, the American president who sought to fight
literal Star Wars with Russia charged the National Commission on
Space with updating the Eisenhower mandate. Ronald Reagan's
committee, which included Gerard O'Neill and Neil Armstrong,
added two major items to the US space agenda. First, building
settlements beyond Earth; second, jump-starting a free market
economy in space. To tie these two imperatives together—territory
and money—the authors called their report *Pioneering the Space
Frontier*.[29]

To make space pioneering seem possible—even necessary—
the committee opened its report with an homage to Christopher
Columbus, whose opening of the New World was a "prelude" to
America's opening outer space. "The promise of virgin lands and
the opportunity to live in freedom brought our ancestors to the

shores of North America," the Committee begins, clearly leaving out the ancestors of those Black Americans who had been forced onto said shores or those Indigenous Americans who were already there.

From the perspective of wealthy white Americans, however, the twentieth-century US exhausted all the "virgin lands" that once secured its "freedom." For this reason, the nation's wealthy white leaders decided they needed another frontier—one that would give Americans room to breathe and, more immediately, to make money. After all, how could anyone live in space without an economy to support them? Therefore, Reagan's committee concluded, "we must stimulate individual initiative and free enterprise in space." The kind that would build the infrastructure to allow territorial expansion onto other planetary bodies, or into Gerard O'Neill's rotating space cylinders.

Economic development will lay the groundwork for the foreign settlements that develop the economy. This was the capitalist promise behind the Virginia Company and the British East India Company, just as it is today for Blue Origin and SpaceX. Private enterprise will pave the way for imperial expansion, which then kicks back favors to the private sector. Hence the recent explosion of corporate space investing. As Texas senator Ted Cruz gushed during a 2018 meeting of the Subcommittee on Space, Science, and Competitiveness, "I predict the first trillionaire will be made in space. I don't know who it will be, and I don't know what they will discover, or what they will accomplish. But I think [space] is every bit as vast and as promising a frontier as the New World was some centuries ago."[30]

Of course, such economic ventures have traditionally depended on the extraction of "resources" and "raw materials" from the "virgin lands" in question. The Americas yielded gold, lumber, ore, and animal flesh along with cotton, rice, tobacco, and sugar; India yielded cotton, silk, and spices, and all of it through the work of enslaved laborers. With this history in mind, one shudders to think what workers and the land itself will undergo with the advent of space mining, the necessary forerunner of any freestanding space economy. We will address the politics and ethics of space mining in the next two chapters. But for now, we'll crawl down our sprawling space beast's last major tentacle: warfare.

For the US to extend its imperial reach across "the last frontier," it will need all of the major forces outlined in the Eisenhower- and Reagan-era reports. It will need the scientific know-how (through NASA and the university system), technological stream-lining (through private industry), significant economic investment (through congressional appropriations and venture capital), and equally significant ideological investment (through appeals to "the American spirit," "Manifest Destiny," and "the salvation of human-ity"). Finally, this high-tech pioneer messianism is going to need some serious military backing. Enter Space Force.

The Infinite Front

In his 2020 State of the Union address—the one that calls it Amer-ica's "manifest destiny" to settle the space "frontier"—President Trump announced the creation of "a brand-new branch of the United States Armed Forces. It's called the Space Force. (Applause.) Very important."

Like most of Trump's adventures off script, this "very impor-tant" riff undermined its own intentions, contributing to a growing snark-heap of jokes, memes, expressions of disbelief, and genuine offense in response to the news that we were getting a space force. The heap included, in no particular order, distress over this decisive effort to militarize space, astonishment at the $738 billion defense bill that created the unit, incomprehension at the president's hav-ing announced the force as a "separate but equal" branch of the military, jokes about the limited heroics of protecting a bunch of satellites, critiques of the force's uniforms (Why would the space cadets need green and tan camouflage in the blackness of space? Would they even *be* in space? Wouldn't they mainly be in window-less cubicles, scrutinizing data streams?), indignation over the logo the Space Force had clearly ripped off from *Star Trek* ("Ahem. We are expecting some royalties from this,"[31] tweeted the former Starfleet helmsman George Takei), guffaws over the revelation that Space Force soldiers would be called "Guardians" (as in "Guardians of the Galaxy?" asked a Twitter-happy Takei[32]); and horror among atheists and Christian pacifists alike over the National Cathedral's having blessed "the official Bible" of this newest warfighting unit at a ceremony in early 2020.

"Almighty God," prayed the Right Reverend Carl Wright, Episcopal Bishop Suffragan for the Armed Services, "look with favor, we pray you, upon the Commander-in-Chief, the 45th President of this great nation, who looked to the heavens and dared to dream of a safer future for all mankind."[33] Both the Military Religious Freedom Foundation and the Anti-Defamation League filed formal complaints over this specifically Christian blessing of an allegedly secular project, while a few beleaguered antiwar Jesus people shook their heads at the sacred endorsement of galactic battle. But considering the half millennium of Christian-military operations in the Americas—or even just considering that Apollo 8 recitation of Genesis—it's hard to be all that surprised by the blessing of an official Bible for the Space Force. ("It's a King James Bible," a colleague pointed out on social media. "So we should expect that soon the Martians will be speaking 17th Century English."[34])

Knowing that it had a public relations problem, the Space Force commissioned a short, elegant recruitment video called "Purpose."[35] Set to sparse piano notes that give way to more urgent strings, the video shows young, focused Black and white servicewomen and men all looking upward—at the stars, immense screens, and towering rockets. A low, calming voice redolent of a young Morgan Freeman invites the viewer to consider their own "purpose" in the face of the "big question" of outer space: "What if?"

The question is deliberately vague, inviting all sorts of imaginings, depending on the kind of recruit you might be. What if our communications network goes down? What if North Korea sends an actual warhead? What if Russia interferes with the actual voting machines this time? What if there's an asteroid coming? What if aliens are watching us? What if they're on their way?

"Maybe you weren't put here just to ask the questions," says the voice. "Maybe *you* were put here [pause] to be the answer. Maybe your purpose on this planet [longer pause] isn't on this planet." The music cuts to silence as the image settles on our blue planet seen from some spacecraft, circled by a long, thin satellite and covered by the Space Force logo.

In the face of multilateral ridicule and anger, the recruiters' strategy in this first video is to appeal to the familiar American theme of sacred destiny. The stars, the serenity, and the language of "purpose" all plug into the viewer's religious imagination, while the lack of specificity (by whom, exactly, were "you put here"?) keeps the religion subtle enough to appease the angry atheists. It's perfect, really—or at least it could have been, if the unit's most lasting lampoon weren't waiting in the wings.

The recruitment video for the United States Space Force was all set to air on May 6, 2020. Suddenly, on May 5, Netflix dropped the trailer for its new, star-studded satire *Space Force*, set to air at the end of the month.

"Our nation's internet runs through our vulnerable space satellites," a fictional secretary of defense announces in the series' first episode. "POTUS wants complete space dominance. Boots on the Moon by 2024. To that end, the president is creating a new branch. Space Force." Before he's done speaking, Mark, the "number 2" air force general played by Steve Carell, snickers loudly until the secretary of defense says, ". . . which Mark will run." Carell's snicker becomes a snort and then an "mmm" before landing on a "whaaaa?"

The show has not found overwhelming popular or critical support, but it did manage to score a second season. At its best moments, *Space Force* intersperses its slapstick antics with genuine political concerns—most notably through the chief scientist played by John Malkovich, who in one episode demands to know why his science budget "pales in comparison to the riches devoted to turning space into an orgy of death."[36] The Carell character calls out the egregious nationalism of the Trump era simply by giving voice to it ("Boots on the Moon!" he tells a quarter-filled auditorium of high school students; "And . . . rest assured these will be US boots. Boots with US feet in 'em").[37]

Cocreated by Carell and Greg Daniels (*Saturday Night Live*, *The Simpsons*, *The Office*), *Space Force* is a critical form of satire familiar to anyone who's watched Jon Stewart, John Oliver, Tina Fey, Kate McKinnon, or Stephen Colbert: more effective as liberal catharsis than social change. After all, *Space Force* has done nothing to undermine the actual Space Force that was born alongside it, a brand-new branch of the US military that the world will now have to live with, and pay for, throughout our generations forever. In fact, *Space Force* seems only to have made the Space Force meaner.

Six months after *Space Force* beat it to the internet, the Space Force released a longer recruitment video, this one with a far less genteel vibe.[38] "Origins" opens with a high-pitched droning, the kind of sound you'd get by running your finger around the rim of a mostly empty wine glass and then hitting it with horror-film distortion. Suddenly, a projectile we can't quite see tears through an orange-red sky and ends in one of those blockbustery *thud*s. The video shifts to another soldier gazing at the sky, but this time the music is ominous and the scene keeps getting interrupted with explosions. Missiles race, canons fire, and rockets flare, shooting massive flames that come out of nowhere, or everywhere, as a clipped, disquieted, exceedingly masculine voice states that "today, space is essential not only to our way of life—it's absolutely critical to the modern way of war."

As you may have noticed, the first video didn't mention war at all. Just purpose and imagination. This second video, by contrast, is filled with references to "enemies," "foreign powers," and reciprocal sabotage. "We will fight in an environment with no up or down," the narrator says, "no left or right, and nowhere to hide." In the hands of the Space Force, warfare is going infinite. (Try to snicker at that, Steve Carrell.)

And in the meantime, the ideology is going incoherent. Having scrapped the sacred destiny line, the recruiters don't seem to have anything up their sleeves apart from a string of oxymorons. The script tumbles into lines calculated to sound pithy that don't actually make sense, like, "we need to stay one step ahead of the future," "the future is where we'll make history," and "we will imagine the unimaginable, anticipate the inconceivable, and prepare for the impossible." These flourishes are all so halfhearted that they're clearly a rhetorical afterthought. What the video is really saying is that the gloves are off and cosmic war is on. No niceties, no higher purposes: just the assertion of national interests in an increasingly weaponized space. Especially now—the voice mentions twice—that space has become the site of such a promising new economy. This is the branch of the military that will secure the asteroidal mines and Martian plantations of the new American frontier.

As the video draws to a close, the explosions get bigger, the scenes cut in faster, and the music gets louder until the sound drops suddenly, the way it did in the first video. Like the first video, this one leaves us with a surveilled Earth emblazoned with a Space Force logo. But while "Purpose" ended in the silence of space, "Origins" ends with that creepy high-pitched drone that haunted its opening. As if to say we're never alone; we're always being seen, always being heard. So if the good guys don't keep watch up there, the bad guys will.

This same bellicose pragmatism runs throughout the force's "inaugural doctrinal manual," published in the summer of 2020 and starkly titled *Spacepower*. Strikingly, a quotation from JFK leads this Trump-era publication: one of those doublespeaky insistences on American dominance for the sake of "freedom and peace."[39] But apart from this quotation (and a saccharine opening line about "ponder[ing] the mysteries of the heavens"), the traditional imperial niceties are gone from this text. The Space Force isn't out there to spread democracy, convert the heathen, or "benefit all mankind"; the Space Force is out there to fight the battles endemic to America's infinite "frontiers." The guardians are not diplomats or internationalists; they are "warfighters who protect, defend, and project spacepower."[40]

From time to time, the authors do concede that one might like to do things in space other than fight. To explore, for example—or to learn, discover, or understand. Certainly, they reason, science is important. But "it is the art of space warfare that gives science its relevance."[41] In other words, science that doesn't serve the military is useless. And at any rate, *Spacepower* concludes, no one will be able to do science in the future at all if we don't "secure the peaceful use of space"—a peace that only "warfare" can secure.[42]

At this point, the space geeks out there (not to mention the international law buffs) might be wondering how the US is getting away with all this "space warfare." Isn't the nation subject to that 1967 UN "Outer Space Treaty," which designates outer space as a realm of "international cooperation" and restricts its "use" to "peaceful purposes"? How peaceful are the Space Force's "key service competencies" of "Orbital Warfare," "Space Electromagnetic Warfare," and "Space Battle Management"?[43] How peaceful is the corporate scramble for contracts, its consequent pollution of Earth's orbit, and its escalating military subcontracting?

Surely, one might expect, the international community has learned something from all these centuries of violent scrambling for resources and land. So what has changed since the Europeans sailed the seas and the white folk pushed their wagons westward? How are nations and corporations expected to behave on the final frontier?

CHAPTER FIVE

Whose Space Is It?

Too much garbage in your face? There's plenty of space out in space!
WALL-E

Leaping for the Species

Neil Armstrong was sure he'd said "a." "*A* man," rather than "man," which would have made no sense. He'd thought about it for months before the flight, finally scrawling the sentence on a piece of scrap paper during a game of Risk with his brother. His brother thought it sounded great. Poor Neil, having to be a poet in addition to a pilot and an engineer and an astronaut.

When the time finally came, the Moon's first earthling tested the ladder's height to make sure he could get back up, then hopped off the landing module's bottom rung. "Armstrong is on the Moon," Walter Cronkite announced.[1] "Neil Armstrong. Thirty-eight-year-old American. Standing on the surface of the Moon. On this July twentieth, nineteen hundred and sixty-nine."

Armstrong interrupts the commentary. Still holding onto the ladder, he moves a moon boot across the "powdery" surface and states haltingly, "That's one small step for man . . . one giant leap for mankind."

Cronkite pauses. His colleague (and former Apollo astronaut) Wally Schirra, confused and lacking Cronkite's vocal majesty, tries to reconstruct what he's just heard. "I think that was Neil's quote," Schirra mutters; "I didn't understand it."

"Uh," Cronkite clamors; "'One small step for man' . . . but I didn't get the second phrase." He asks for help from one of the network's monitors—maybe someone stationed in Houston—and a few moments later a more confident Uncle Walter comes back to repeat the instant proverb, which still doesn't make that much sense. If it's a small step for "man," which is an old-school masculinist way to say "humanity," then how can it also be a giant leap for "mankind," another old-school masculinist way to say "humanity"? Without the "a," it seems like Armstrong was saying, "that's one small step for man, one giant leap for . . . man."

After he landed, the Moonwalker said he'd certainly meant to say "a." It was a small step for *him.* One man. Neil, clutching the ladder in his big white space suit on the surface of Earth's constant companion. Meanwhile, this "small step" was a "giant leap" for the rest of us, who had suddenly become the kind of creatures who could walk on another world.

As we have already seen, this "for all mankind"-ism serves as the running ideological theme of the US space program. It shows up in Eisenhower's first post-Sputnik statement, the Episcopal blessing of the Space Force Bible, and that plaque Apollo 11 left on the Moon: "We came in peace for all mankind." Military aspirations, economic possibilities, and political venom aside, the message the US wants to send the rest of the world is, "we did this for you."

In the months before Apollo 11, NASA appointed a Committee on Symbolic Activities for the First Lunar Landing. Yes, you read that right: there was a whole group of people working on the ritual details of the mission. An old memo shows that Committee Chair Willis Shapley was concerned above all to strike a balance between American nationalism and global humanitarianism. "The intended overall impression of the symbolic activities," he wrote, "[should be] of an historic step forward for all mankind that has been accomplished by the United States of America."[2] And there's that doublespeak again. We did it; we did it first, and we did it first *for you.* (You're welcome.)

This was the same committee that directed the Apollo 11 crew to plant the American flag on the Moon. Considering the number of international citizens who had contributed materially, scientifically, and operationally to the mission—and considering the benefit Apollo was allegedly rendering to "all mankind"—the committee had considered installing a United Nations flag in the lunar dust, but ultimately decided to temper its internationalism with a celestial dose of Old Glory. Congress agreed with the Committee on Symbolic Activities, revising the NASA appropriations bill to insist that any mission exclusively funded by the US would refrain from flying the flag of any other nation, or international body, on the surface of the Moon.[3]

What, then, does "for all mankind" mean? In the minds of folks like Disney and von Braun, it carries that dubious species-level freight, implying that the same creature who began in caves, invented tools, and harnessed wind, steam, and electricity is now making its next evolutionary leap from the rocks to the stars. But the usefulness of this grand story is questionable to anyone with more particular concerns than the alleged history of the species. How would the late-sixties lunar landing advance the civil rights of Black Americans back on Earth? How would it contribute to resolving the mess Apollo was escaping in Vietnam? How would the moon shot help decolonize India and Africa? What was its stance on the labor movement, the women's movement, reproductive justice, gay rights, food shortages, poverty, dictatorial regimes, refugee resettlement, nuclear proliferation, water rights, and the growing sense that there was something very wrong with the climate?

The jazz poet Gil Scott-Heron captures this dizzying imbalance between the moon shot and nearly every other human concern in his spoken-word piece "Whitey's on the Moon." In the particular split screen of this poem, the speaker and his family are struggling with medical crises, financial shortfalls, canceled utilities, and basic sanitation problems, "while Whitey's on the Moon." As the poem progresses, we realize it's not just a coincidence that white folks are up in space while Black America is falling apart, nor is it just political sleight of hand or Earth-hating escapism (à la Bezos and Branson on their cosmic joyrides).

As Scott-Heron's speaker explains, he is paying too much in rent because his landlord is paying too much in taxes because "Whitey's on the Moon." Of course, the speaker is paying taxes, too—nearly his whole paycheck. Meanwhile,

> The price of food is going up—
> And as if all that shit wasn't enough,
>
> A rat done bit my sister Nell
> with Whitey on the Moon.
> Her face and arms began to swell
> but Whitey's on the Moon.
>
> Was all that money I made last year
> for Whitey on the Moon?
> How come there ain't no money here?
> Hmm! Whitey's on the Moon.

So it's not just that the Apollo astronauts happened to be on the Moon while poorer, darker-skinned Americans were struggling with racism, poverty, and disease. It's not just that the Apollo missions intentionally distracted Americans from the convulsions of protracted war and anti-Black violence. It's not even just that the missions squandered the money that poor and historically oppressed people so desperately needed (this was the critique of Martin Luther King's successor Ralph Abernathy, who led five hundred people, two mules, and a wooden wagon to the Kennedy Center to protest the moon shot). It's that this race to conquer the heavens actually depended on the American "many," whose degradation fueled the elevation of the American "few" who then claimed they were conquering the universe for all of humanity.

In today's register, the problem isn't just that the $1.5 billion that Bezos spends every year and a half on Blue Origin would deliver lead-free water to the residents of Flint, Michigan. It's not just that Musk's antics in space distract us from the "unhealthy and unsafe" working conditions of SpaceX.[4] It's that US workers in and beyond Flint and Boca Chica are paying the taxes that fund the national space program that's giving itself over to the private sector. Meanwhile, the new astropreneurs not only dodge the taxes that poorer folks pay; they then receive those taxes *themselves* in the form of federal grants and contracts. Honestly, it's as if Americans are paying taxes directly to Bezos and Musk.

And predictably, the spacemen justify the whole enterprise by saying they're saving the species. The forests blaze and oceans rise and Whitey's on the Moon. Even worse, Whitey says he's on the Moon "for all mankind."

Naive as Armstrong and his colleagues may have been about the social importance of the Moonwalk, a corporate orgy on a burning planet couldn't possibly be the historic "leap" he had in mind. But again, what *did* he have in mind—along with NASA's ritual committee, Eisenhower, Johnson, Kennedy, and Nixon? Was the "giant leap for mankind" strictly symbolic? Or did the US imagine it would actually bring material benefit or psychic uplift to the people of the world? In what sense was the lunar landing "for" the rest of us?

The key to answering this question, and to understanding how the promise of 1969 became the reality of the 2020s, is to get clear about the recurring preposition. Whom is outer space really for? And what does anybody mean by *for*?

Common Space

Two years after the world-shaking launch of Sputnik, the United Nations established a Committee on the Peaceful Uses of Outer

Space (COPUOS) "to govern the exploration and use of space *for the benefit of all humanity*." In this mission statement, the US slogan gained international reach and offered some a genuinely global promise. If the space race was really "for" all humanity, then a representative body would assemble itself to determine and adjudicate the scope of that *for*.

Since 1961, COPUOS has been organized into two branches. The Scientific and Technical Subcommittee assesses the range, promises, and dangers of contemporary uses of space—from meteorology and navigation to communications and education, agriculture and health care, military operations and national defense. Meanwhile, the Legal Subcommittee discusses liability, cooperation, and above all, property.

The question of property may bring to mind the curious case of Dennis Hope who, you might recall, sells parts of the Moon (and the rest of the solar system) from an office in Northern California. Hope insists he is able to sell this property because he owns it, having found a loophole in the 1967 Outer Space Treaty. Technically called the "Treaty on Principles Governing the Activities of States in the Exploration and Use of Outer Space, Including the Moon and Other Celestial Bodies," the Outer Space Treaty (OST) is the first and most powerful of five treaties the UN has ratified in the area of international space law. And Article 2 of the OST says quite clearly that "outer space, including the Moon and other celestial bodies, *is not subject to national appropriation*." That means that no one can own it.

"No," Dennis Hope might say; "that means that no *nation* can own it. The treaty doesn't mention individuals, so I, Dennis Hope, have claimed the Moon and other celestial bodies, which I intend to keep selling to the good people of Earth." Unfortunately, the treaty's failure to mention individuals—or corporations, which under US law maintain the status of legal personhood—is just one of its numerous shortcomings. And yet it's the treaty we've got.

As the UN faced the question of space property from the late nineteen fifties into the sixties, it was well aware of the massive inequalities dividing the former colonial powers from the nations they'd exploited to industrialize themselves. The poorer countries, along with some of their chastened wealthy allies, were worried that the "new frontier" of space would exacerbate the inequalities of the earthly one if it weren't governed by a different set of rules. The earthly frontier, we'll recall, was guided by the principle of *terra nullius,* or no one's land. If a territory could be said to be empty, or underused, then it was subject to seizure or purchase by the first European nation to claim it.

Clearly, then, when the OST says that outer space "is not subject to national appropriation" by any means, it is trying to set outer space on a new, enlightened course. It is trying to say that we're playing a different game from the frenetic land grab of terrestrial modernity. Space, the authors wanted to say, is not what political theorists call a *res nullius* (empty thing); it is a *res communis* (common thing). So it's not out there for the most powerful nations to claim; it's out there *for* all of us.

This communal intention is clear from the treaty's preamble, which professes a "belief" in the use of outer space "for the benefit of all peoples irrespective of the degree of their economic or scientific development." This communalism is also clear from the first article, which declares that space shall be used and explored "for the benefit and in the interests of all countries" and which calls space "the province of all mankind." Unfortunately, the scope of this province and the power of the commons is dramatically curtailed by the very next sentence, which declares outer space to be "free for exploration and use by all States."

At first glance, this declaration may seem to share the beneficence of the language leading up to it: outer space is for everyone, therefore "all States" can "explore and use" it. But how exactly are "all States" supposed to get there? At the time the OST was signed in 1967, the only nations with satellites were the USSR, the US, and France. Only a few more had even initiated a space program. Many had only recently won their independence from imperial powers who'd spent the last century stripping their lands of resources and running off with the profits. What good does it do to tell a nation it is free to go to space if it doesn't have the means of going?

So there's a profound tension in the OST between equality and freedom. The principle of equality says that everyone has the same right as everyone else to outer space. The principle of freedom says that everyone is permitted to pursue it. But since very few nations *can* pursue it, the freedom undercuts the equality. The rich nations rush off to space, reading the Bible and planting flags "for all mankind," while the poor nations fall farther behind.

There are a number of provisions in the OST, the details of which would likely bore both you and me to tears. Many of these provisions are expanded in three subsequent treaties, which commit signatories to rescuing stranded astronauts, assuming liability for the things they put in orbit, and registering those objects publicly. But for all these ancillary stipulations, there are two major declarations in the OST. The first is that outer space is "for the benefit of all peoples," and the second is that space must only be used "for peaceful purposes."

We have already seen the egalitarianism of the "benefit of all peoples" clause weaken under the "freedom" that tramples it. And unfortunately, a similar fate befalls the "peaceful purposes" clause.

In order to avoid the disastrous wars that precipitated the UN's formation, Article 4 of the OST states that nations may not put nuclear weapons or weapons of mass destruction in space. Such arms may neither be installed on celestial bodies nor placed in orbit nor stationed "in outer space in any other manner." Rather, outer space must be used "exclusively for peaceful purposes." Just as it did in Article 2, however, this communal declaration in Article 4 gives way to a massive exception. And then to another right after it:

1. "The establishment of military bases, installations and fortifications, the testing of any type of weapons and the conduct of military manoeuvres *on celestial bodies* shall be forbidden" (emphasis added).

2. "The use of military personnel for scientific research or for any other peaceful purposes shall not be prohibited."

Rereading that first sentence, we might notice that although it forbids the building of bases and testing of weapons—even conventional weapons—"on celestial bodies," it does not forbid engaging in such activities in orbit or on transiting spacecraft. And although the first sentence prohibits the installation of military *bases* on the Moon or other celestial bodies, it condones the installation of military *people* on such bodies. As the second sentence confirms, a nation can station all manner of soldiers in space so long as said soldiers are pursuing "peaceful purposes."

It is this gargantuan loophole that the Trump administration will exploit with its introduction of the sixth branch of the US military. One might think it absurd, really, that any nation publicly dedicated to the OST's "peaceful purposes" could even *think* of establishing a space force. Indeed, many UN member states have been baffled and dismayed by what they see as a clear act of aggression on the part of the US.[5] And yet the US insists that it is behaving defensively rather than aggressively. Both Russia and China have already consolidated their own space operations into one military branch; they are both developing antisatellite technologies; and China has expressed its intentions to "colonize the solar system and beyond." So as far as the Pentagon is concerned, the Space Force is an assertion of national self-defense, guaranteed by Article 51 of the UN charter.[6]

Even if it adheres to the letter of the law, though—and I'm not sure it does—the Space Force misses its spirit entirely. The OST commends outer space to the pursuit of "peaceful purposes," while the Space Force is touting space as "the world's new war-fighting domain."[7] More strangely, it does so while declaring fidelity to the OST. In fact, one report argues, the Space Force will *enable* US adherence to the OST in the face of their "potential adversaries'" developments. Russian and Chinese antisatellite programs are threatening to interfere with US operations, especially in the growing sector of the space economy, and yet the OST guarantees the right "to explore and to use space for peaceful purposes." How can the US enjoy the "free use" the OST promises if foreign powers can jam its satellites and scramble its signals? In short, without a space force, the US will be unable to exercise its right of "unfettered access" toward the (peaceful?) pursuit of financial gain.[8] As the force's doctrinal manual explains, "the success of these endeavors is only possible if we secure the peaceful use of space," and they can only secure such peace through war.[9]

Uncommon Space

Although the sixth branch of the US military and the astropreneurial crusade are both recent phenomena, the trends toward militarization and appropriation became clear the minute Armstrong and Aldrin planted that flag on the Moon. Realizing that the OST was insufficient to prevent outer space from becoming another theater of exploitation and war, COPUOS spent the 1970s drawing up the "Moon Treaty" to fill in some of the gaps in the OST.

Formally named "Agreement Governing the Activities of States on the Moon and Other Celestial Bodies," the Moon Treaty opens by stating its parties' desire "to prevent the Moon from becoming an area of international conflict."[10] Toward that end, the signatories agree to many behaviors familiar from the OST. These include reserving space for peaceful purposes, refraining from placing nuclear weapons in space, and pursuing space exploration "for the benefit and in the interests of all countries, irrespective of their economic development." Unlike the OST, however, the Moon Treaty goes on to explain what "for the benefit of all" *means*. And the meaning is not just symbolic.

As the Moon Treaty's Article 11 declares, "The Moon and its resources are the *common heritage of mankind*" (emphasis added). This phrase was the coinage of Maltese ambassador Arvid Pardo during the 1967 discussion of what would become the UN Convention on the Law of the Seas. Seeking to avoid the "competitive scramble" that could only make "the strong stronger, the rich richer," Pardo suggested that the deep seabed and ocean floor be understood as "the common heritage of mankind." What he meant was that any resources retrieved from international waters would be shared among all nations. This isn't the way things ultimately unfolded in the deep seas, but in the meantime, the originally communal intention of the Law of the Seas made its way into the Moon Treaty.

Intensifying the OST, the Moon Treaty prevents nations from appropriating celestial bodies, parts thereof, or "natural resources in place." So no nation can own the Moon, regions of the Moon, or the ice (for instance) on its polar caps. As such, any resources extracted from the Moon or other celestial bodies are subject to "an equitable sharing by all States Parties," especially those states in the "developing" world. In other words, the rich countries have to share at least some of what they take with the poor countries, even if the rich countries run the mission "themselves" (after all, who provided the centuries of resources and forced labor that made the rich countries so rich?). Finally, Article 11 says that "as soon as such exploitation is about to become feasible," the parties will establish an international body to regulate and govern the commercial extraction of lunar materials.

To my ears, at least, the Moon Treaty sounds pretty reasonable. The spacefaring nations, and the US in particular, keep saying they are exploring space "for all mankind." The nations have agreed by means of the OST that space should be reserved for peaceful purposes. So to build a peaceful community for "all," to avoid a galactic rerun of earthly imperialism and maybe even repair some of its damage, we'll set up an international body to allocate resources and make sure everyone benefits from whatever it is space has to offer (they weren't quite sure at the time). This way, that small step for Armstrong might actually become a giant leap for humanity.

Except the US wouldn't ratify it. Neither would Russia or China or most of the UN's other member states. By 1984, the treaty had gained just enough support to become international law, but that only gave it standing among its nine signatories, none of which was a spacefaring nation.

This story could have unfolded otherwise. The US delegation did, in fact, sign the Moon Treaty, but the Senate refused to ratify it. Why did it refuse? As Scott Pace, executive secretary of Trump's National Space Council, boasted in 2020, Pace established a "grass-roots campaign" in 1979 "to ensure that the United States would not sign or ratify the . . . Moon Agreement." Refusing even to call it a treaty, Pace explained that his opposition rested on two main convictions: first, that US actors should not be beholden to any unelected, international regulatory body (like the one proposed in the Moon Treaty), and second, that private investors would not back a mission to recover space resources unless they were guaranteed exclusive ownership of them. In Pace's view, anyone who was paying attention could tell that outer space was about to explode economically, and the Moon Treaty would be bad for business. It may come as no surprise that Pace was an early member of The L5 Society, dedicated to realizing Gerard O'Neill's vision of rotating orbital space colonies.

This demolition of the Moon Treaty set the US on the path that led to Ronald Reagan's dreams of a cosmic gold rush, George W. Bush's attempt to privatize space exploration, and finally Barack Obama's Commercial Space Launch Competitiveness Act (CSLCA). As you may recall from chapter 1, this is the energetically bipartisan 2015 law that declares space resources to be the property of any "US citizen" who manages to extract them. Not the "common heritage of mankind"; just the uncommon heritage of whoever's rich enough to get up there and grab what they can.

This "whoever" includes corporations, which the US Supreme Court determined in 2010 to be "people" under the law. So by all accounts, the 2015 CSLCA gave investors the last bit of confidence they needed to pour their money into space. In 2017 alone, prospectors funneled nearly $4 billion into commercial space ventures, a number that amounted to nearly half of all private investments in all industries over the preceding five years.[11] At the time of my writing, the global space industry is worth $350 billion and is projected to balloon to more than $1 trillion by 2040.[12] Facing this coming windfall, one space-mining CEO went so far as to call the CSLCA the commercial equivalent of the Homestead Act.[13] Finally, the final frontier was open.

In early April of 2020, I heard through my space-justicey social media channels that Donald Trump had unilaterally handed over space "resources" to private ownership. In the language of Executive Order 13914, "the United States does not view [outer space] as a commons. Accordingly, it shall be the policy of the United States to encourage . . . the public and private recovery and use of resources in outer space."[14] Doomscrolling from one news outlet to the next, I was horrified. Isn't the US a party to the OST, which calls space "the province of all mankind"? Hasn't every president since Sputnik assured us that "all humanity" would "benefit" from US leadership in space? Didn't Neil say he was Moonwalking for all of us?

What I've come to realize, however, is that Trump's executive order doesn't actually do anything new. It presents the US space program as it's always been, just without its traditionally humanitarian coating. From Johnson's quest for "total control" to Kennedy's insistence that "we must be first" to the ritual banning of international flags to Obama's corporate space act, the US position has always been, as Moon Treaty–killer Scott Pace is happy to declare—without philanthropic flourish—that "outer space is not a 'global commons,' not the 'common heritage of mankind,' not a *res communis*,' nor is it a public good."[15]

Of course, there are people who disagree. In addition to the growing number of Indigenous, Black, Western, and Global South scholars and activists who will guide our journey over the next two chapters, there are still voices within the UN insisting that space be understood as a commons. Even at the 2021 meeting of COPUOS, there were delegations pleading that the UN not promote "the commercialization of space," since space is "the common heritage of all mankind."[16] Even then, there were calls for the spacefaring nations to stop militarizing the heavens, to stop circumventing the UN to create their own laws and treaties, and to sign the Moon Treaty so everyone could do "space" differently. But as it turns out, COPUOS has no juridical power. Member states might charge one another with having violated the OST, but there isn't much anyone can do about it. So if Russia and China and the US have decided space isn't a commons, then for all practical purposes, it's not.

And since "practical purposes" are all that really count, NASA decided to *prove* space isn't a commons by buying some of it. In September 2021, the space agency paid Lunar Outpost ten cents as a down payment for some lunar soil. Once the space-mining company gathers the regolith and deposits it elsewhere on the Moon, NASA will pay the firm ninety more cents for the "delivered" materials. As NASA explains, "this process will establish a critical precedent that lunar resources can be extracted and purchased from the private sector in compliance with Article 2 and other provisions of the Outer Space Treaty."[17] In other words, buying lunar resources will demonstrate that it's possible to buy lunar resources. And if anyone objects, they can bring it to COPUOS, which will dutifully record the objection in minutes that nobody reads.

The Circle Game

For the US in particular, the reason space can't be a "common heritage" is that the logic of the commons undercuts the logic of capitalism. If you're sharing resources, you can't maximize profits. And maximizing profits has become the central aim of the new corporate space race, not just in the minds of America's astropreneurs but according to the official priorities of the nation itself.

According to the 2020 Space Policy of the United States, the first goal of the space program is to "stimulate economic growth"[18] The last is to "advance economic freedom." There are a few others, like improving quality of life and spreading democracy, but these are sandwiched between expanding the market (economic growth) and deregulating it (economic freedom). Similarly, the Department of Defense's 2018 "Space Force Report" lists three major missions, the first of which is to "protect our economy."[19]

In his "Introduction to Outer Space" for the American people, we might remember that Dwight Eisenhower listed four reasons it was crucial to pursue a space program. These were: exploration, defense, "national prestige," and science.[20] In the work of Eisenhower's presidential descendants, a fifth reason has emerged to rule them all: money. Even the Department of Defense admits that its first priority is securing the national economy. Meanwhile, "national prestige" rests on economic power, "exploration" means resource hunting, and "science" . . . well, science doesn't tend to get mentioned, except as a kind of handmaiden to the technology enabling the new economy.

That may be a bit of an overstatement. But not by much. Scientific priorities do come to voice in contemporary NASA publications, but such priorities are almost always explained as the means toward military-economic ends—specifically, the ends of heading back to the Moon and then advancing to Mars.

The Artemis Program, we may recall, is NASA's response to the Trump-Pence directive-threat to have "boots on the Moon" by 2024. As NASA explains at the top of its website, Artemis is named after the twin sister of Apollo because the mission will "land the first woman and first person of color on the Moon."[21] (It's a racially attuned revision of the earlier language of the Trump administration, which had demanded that 2024 see "the next man and the first woman" walking on the lunar surface.) As NASA summarizes the mission, this first woman and first person of color—who might well be the same astronaut—will work with "commercial and international partners" to "establish sustainable exploration" and then "use what we learn . . . to take the next giant leap—sending astronauts to Mars."

The timeline is ambitious. Remember that 2019 speech, when Mike Pence told NASA to "step up" and get the job done, otherwise he'd find someone else? Biden officials have hinted that they're likely to give NASA officials more leeway than their predecessors did. As far as the stated mission is concerned, however, the aim is still to return astronauts to the Moon by 2024, establish a permanent outpost by 2028, and head to Mars sometime in the late 2030s.

The question one might ask is, Why? Why travel back to the Moon rather than to some other cosmic body? Why head to Mars rather than, say, Venus, which until very recently spent decades going unfunded and unexplored?[22] Why staff Artemis with human astronauts rather than just rovers and robots? And while we're asking, Why do any of it? Why not invest in those space technologies—like weather tracking, energy efficiency, disaster relief, and environmental protection—that directly benefit Earth?

I hate to tell you this, but I haven't found a clear answer. What I've found instead is a logical circle: we are establishing a long-term presence in space to retrieve and use the resources that will establish a long-term presence in space. We need the colony to anchor the economy and need the economy to sustain the colony. But why do any of it at all?

Even as they're working toward similar goals, the private and public sectors give different answers. The corporate messiahs and their followers warn of imminent extinction: if we don't expand into space, they cry, the species will never survive. Meanwhile, NASA remixes JFK soundbites about "American leadership in space." In the words of NASA's 2020 Artemis Plan, "the Moon to Mars approach will assure that America remains at the forefront of exploration and discovery."[23] The stringent timeline will assure us we get it done as quickly as possible (faster, preferably). But why does America need to remain at this particular "forefront"? Whence the urgency?

The NASA authors seem to assume it's self-evident: we've got to get to Mars, and fast. So rather than giving an answer, they dust off a few more rhetorical gems and say the mission will "engage and inspire America and the world" without explaining what that means.[24] They call Artemis *"humanity's* quest to create a future," having just declared they're seeking US dominance over the rest of humanity. And they extol the "endless discovery and growth in the final frontier," as if the image of heroic pioneers will imbue the mission with automatic importance and cause us to forget what we were asking. But if we shake off the romance and nostalgia, we can stick with the question: *What are we doing this for?*

Perhaps surprisingly, the most honest answer comes from Jeff Bezos. It's not that we can't live unless we go to space; it's that we can't live *like this*. The capitalist economy that fueled the Industrial Revolution that fired out the Digital Age has depended since the late fifteenth century on the extraction of resources and the exploitation of labor. And although there unfortunately seems to be no end to the forms that slavery and indenture can assume in the modern world, the profiteering that relies on such labor is coming up against real material limits. There is only so much oil in the ground, gold in the hills, gas in the mountains, clean water in the lakes, and titanium in the mines.

In this context, the promise of deep space is the promise of infinite resources. According to one estimate, the asteroid belt alone contains metals that could provide "$100 billion for every person on earth."[25] Of course, they're never going to. Thanks in part to the demise of the Moon Treaty, "humanity" will not reap the benefits of the burgeoning space economy any more than the people of Sierra Leone have reaped the benefits of the diamond trade. Moreover, it would be prohibitively expensive to bring the water, ore, gold, and platinum buried in asteroids back to Earth. Even if you could just "beam them down," any massive influx of asteroid nickel (for example) would tank the market in terrestrial nickel. So, with the exception of some of the rare-earth elements that manufacturers use in very small quantities, the resources recovered from space will mostly be used in situ. The water on the Moon can refuel rockets. The heavier elements will help construct tools, bases, vehicles, and habitats without our having to schlep the heavy materials all the way from Earth.

Welcome to the circle game. Why are we mining outer space? So we can live and work and explore there. Why are we living and working and exploring outer space? So we can figure out how to mine it.

What we're caught in here is "the cyclical logic of capitalist growth" that tells us "we must expand, in order to keep expanding."[26] If the market doesn't grow, then profits plummet, and the market simply can't grow infinitely ... not, at least, on a finite planet. This seems to be what Ayn Rand understood when she ended *Atlas Shrugged* with the libertarian hero John Galt raising his hand, "and over the desolate earth he traced in space the sign of the dollar."[27]

For Galt's contemporary devotees, the race to space is the only way to sustain what climate activist Greta Thunberg has called late capitalism's "fairy tales of eternal economic growth."[28] Fairy tales like the one that assures Bezos his grandchildren should be using more energy than he does. More energy to power more devices to trade more shares of more companies that buy and sell the resources that make more devices. Devices powerful enough to video chat the earthbound remnant from your rotating pod out in space.

Bounded in a Nutshell

To Bezos, of course, it's not a fairy tale. We really will live in space colonies—at least, we *should*, if we want to have any sort of future other than a "dull" one. As far as he can see, infinite growth is only impossible given finite resources. But the moment we unlock the infinite resources of the infinite universe (here come the schmaltzy promises), there's nothing we can't do, no goal we can't reach, no limit to our dreams.

The first part of the conclusion may well be true: the universe does seem to be infinite. Granted, it may be finite ("But," as my littlest brother asked me when he was eleven, "where would it end?").[29] Whether it's bounded or boundless, however, the universe is so enormous that there's honestly no reason *not* to call it infinite—at least from the perspective of our comparatively minuscule planet, solar system, and even galaxy. That's not the problem with Bezos's reasoning.

The problem is that, like Musk, Bezos says that his adventures are making this infinite space *accessible* to "humanity." They're not. They're making profits for a very small cadre of wealthy folks by means of a powerful myth. Meanwhile, the rest of humanity scrambles to make rent, find clean water, pay for health care, evacuate before the fires and floods hit their houses, survive a traffic stop or border crossing, or hold onto a job and an unvaccinated baby at the same time. Taking this system and extending it to Mars, even if it were possible, wouldn't open "our" dreams to infinity. Rather, it would keep us stuck in the same violences, fears, and inequalities that the extraction-and-settlement model has produced on this planet. Except on Mars, we'd have to pay for air.

The clearest indication of the limits of "infinite space" is the growing pile of garbage around us. Not just in the oceans and overstuffed landfills, but literally around us: in orbit, where deep space is decidedly finite.

* * *

A long, long time ago, before anything crawled on Earth or flew in its skies or swam in its oceans, an enormous *something or other* smashed into our molten planet, causing bits of it to fly off into orbit. As these bits cooled and congealed, they formed our Moon. Big enough to stabilize Earth's wonky orbit and calm down its climactic convulsions, this Moon rules the tides that produced the bacteria that produced the invertebrates and then the vertebrates, the reptiles and the mammals and the plants and the primates . . . you know the rest. For four and a half billion years, Earth's only satellite was its Moon. And then in October of 1957, the Moon got company. Sputnik and its carrier rocket would now join the Moon in its lonely ovals.

Since then, the circuit has grown a lot more crowded. For decades, the spacefaring nations thought very little about leaving their trash out in orbit; after all, space is so big! Unfortunately, the waste of overdeveloped nations seems to be even bigger. Sixty-five years after Sputnik, there are over seven thousand satellites in orbit,

more than half of them defunct, and all of them traveling at eighteen thousand miles per hour.[30] There's other stuff as well: tools
that astronauts have accidentally lost on spacewalks, urine they've
put there on purpose, old rocket stages, bolts and brackets and
screws, and the ashes of Gene Roddenberry—along with anybody
else who's paid a private company to fire their remains into orbit.[31]

You may be asking, "Doesn't anybody clean up around there?"
It's a great question. You and I can't go to a beach or state park without carrying out what we carry in. But space travelers can. The six
Apollo missions left the surface of the Moon littered with a total of
ninety-six bags of human waste—urine, feces, vomit, and bits of
food.[32] They did it to offset the weight of the lunar rocks they were
bringing home. They did it because they thought the solar radiation
would "sanitize" the ninety-six bags of waste. But surely they knew
the bags themselves, and the diapers and wipes they contained,
would still be there? (Along with the landers and the flags, that
strange little plaque, and the Bible an astronaut left on a rover?[33])

As gross as it is on the Moon, the garbage problem is far worse
in orbit. Even after it became clear that there was just too much
stuff hurtling around up there, it did not become clear what anyone
could do about it. Ideally, defunct machinery will fall into Earth's
atmosphere and get incinerated. Some of it will survive the inferno
and crash into our oceans, backyards, forests, and Indigenous land.
But most of it just stays its careening course forever, headed at
breakneck speeds for everything else out there.

Unsurprisingly, this hurtling space crap does a lot of damage.
Even tiny pieces of metal can carve divots and holes in the Hubble
Space Telescope and crack the glass on the International Space Station. Larger objects could disable the ISS completely. And when
two big things collide—the way numerous defunct satellites have
done—they send thousands more pieces of shrapnel in every direction to do even more damage. If enough of these events pile up,

collisions would trigger one another in an unstoppable cascade.
The fragments would grow smaller, more numerous, more uni-

form in direction, resembling a maelstrom of sand—a nightmare scenario that became known as the Kessler syndrome. At some point, the process would render all of near-Earth space unusable. Theoretically . . . our planet could acquire a ring akin to Saturn's, but made of garbage.[34]

There are a few cleanup proposals out there, the most promising of which is a cube-shaped satellite that aims to chase space garbage with a harpoon and net.[35] But this mash-up of Wall-E and Moby-Dick hasn't yet managed to spear anything. And even when it does, it's not clear who would actually pay for it. Which nation is going to foot the multimillion-dollar bill for scrubbing some rubbish out of orbit? Which star-addled private investors will fund the R&D for a cosmic janitorial system?

As of 2021, NASA is tracking over twenty-six objects swirling around Earth and endangering the agency's missions. But that's just the big stuff. The European Space Agency currently estimates that the skies are riddled with thirty-four thousand objects greater than ten centimeters, nine hundred thousand objects between one and ten centimeters, and 128 million objects between one millimeter and one centimeter.[36] And 95 percent of it is garbage.[37]

There is no globally recognized catalog of all this debris and therefore no reliable way to avoid colliding with the small stuff, but one aerospace engineer and self-proclaimed "space environmentalist" at the University of Texas is trying to create one. Moriba Jah's AstriaGraph tracks and monitors space debris by relying on crowd-sourced data.[38] Ultimately, AstriaGraph aims to be "the equivalent of a Waze app for space traffic," with everyone from citizen-observers to national space agencies dropping pins to warn fellow spacefarers of incoming rocket parts and sprocket kits.[39]

Meanwhile, the problem's getting worse. Jah's images get denser every two weeks, when SpaceX delivers another sixty Starlink satellites into orbit. Having already launched over fourteen hundred of them around Earth, the company aims to add another forty-two thousand by mid-2027. In the meantime, Blue Origin has gained

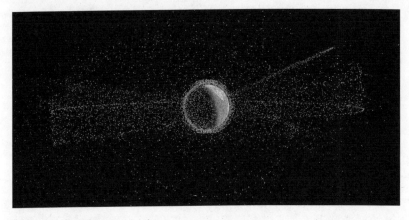

Figure 5.1 Orbital space debris.

ASTRIAGraph: Computational Astronautical Sciences and Technologies group, Oden Institute for Computational Engineering and Sciences, University of Texas at Austin

clearance to launch a few thousand satellites of its own.[40] Musk tends to boast that while Blue Origin's satellites are "at best several years from operation," his own are on the verge of global coverage.

Well, they're on the verge of *potential* global coverage. SpaceX is throwing all these satellites into orbit before the company has even got a full-fledged internet service. What they've got is a beta test called "Better Than Nothing" that provides internet to subscribers for $99 a month, plus $499 for a heavily subsidized piece of satellite hardware named "Dishy McFlatface." Musk is really hoping he doesn't lose so much money on the venture that Starlink has to close up shop, like every other satellite-based telecommunications company before it. But whether Starlink makes it or not, most of the satellites are up there for good.

What's more, each of these satellites and pieces of shrapnel reflects bits of sunlight that mar astronomical observations with screaming white streaks. Astronomers estimate that the last decade of energetic launches has brightened the night sky by 10 percent, even in places with no terrestrial light pollution.[41]

To its credit, SpaceX has responded to the astronomers' distress by hitting the satellites with some black coating so they reflect less light. It's still not enough to abate the light pollution, though. Plus, the coating carries the risk of overheating the machines and causing them to malfunction. So Musk does what every superpower does when it's ravaging an environment: he talks about all the good he's doing for "humanity." Just as pesticide corporations raged against Rachel Carson's *Silent Spring* in the early 1960s by saying their toxic pesticides were feeding the globe,[42] Elon Musk proclaims his satellites will provide digital salvation to the analog masses. The Starlink "constellation" will reach the people ordinary tech has left behind: rural Americans, villagers in Africa, Indigenous folks around the world. Starlink, like the five hundred years of colonial ventures that produced it, proclaims itself an act of *service*.

Many folks aren't so sure. In its beta stage, Starlink's only subscribers are middle- and upper-class folks who can afford a $500 piece of equipment and an extra hundred in monthly bills. Even if

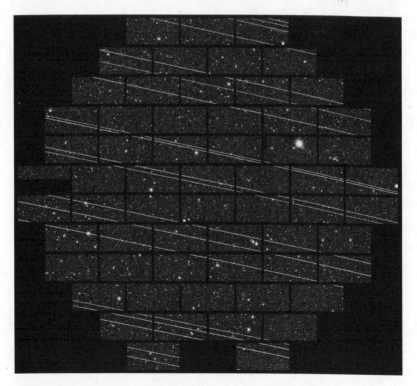

Figure 5.2 Streaks from nineteen Starlink satellites across the night sky.

CTIO/NOIRLab/NSF/AURA/DECam DELVE Survey

prices go down, a recent study has shown that satellite-based internet will still be far more expensive than ground-based internet and that it won't work well in high-density areas.[43] In the meantime, African space agencies would rather put their own satellites in orbit than rely on the whims of wealthy cowboys.[44] And as for the "service" these satellites will render to Native peoples, pushed over centuries to lands at the edges of the terrestrial grid, Moriba Jah has summarized the problem quite clearly:

> No one asked regular human beings what they thought about [satellite constellations]. Nobody went to Indigenous people who have very dark skies and have been using the skies for a long time to help them understand their relationship to seasons. . . . Their skies are now changed forever. And nobody asked them.[45]

At the beginning of this book, I said that space was a mess, but it's worse than that. Space is a *disaster*. The word literally means that the stars (*astra*) are out of place, throwing everything out of order. And like any genuine disaster, the disaster of space is affecting everyone. Indigenous navigators can't see the stars. Western astronomers can't get decent images. The International Space Station has to perform heroic maneuvers to dodge incoming projectiles. And investors worry about their satellites getting demolished in a high-speed garbage race. Even the Space Force admits that orbital debris is becoming a significant impediment to their operations. (Lest they be further ridiculed for functioning as a sanitation unit, however, the Space Force is requesting that some civilian agency figure out how to take out the trash.[46]) In short, everyone has a shared interest in not getting pummeled by debris. Could this shared disaster finally make it clear to the corporate cheerleaders and cosmic nationalists that space is a commons after all? Might the space junk threatening to strangle us actually be our salvation?

CHAPTER SIX

The Rights of Rocks

Well, the mountain was so beautiful that this guy built a mall and a pizza shack.
DAR WILLIAMS

Inhuman Resources

It was the mid-1980s. Thatcherism gripped the UK, the USSR turned toward private enterprise under a centrist Gorbachev, and Reaganomics promised US workers a trickle that never made it down. "Born in the U.S.A." scored the summer of 1984 in all its post-Fordist, post-Vietnam futility. "Everybody Wants to Rule the World" lent the summer of 1985 its moody resignation to market economics, stalemated superpowers, and escalating climate change. And in Princeton, New Jersey, the physicist-prophet Gerard O'Neill was teaching the world's future rulers how to expand their empires into space.

Although he studied engineering and computer science rather than physics, Jeff Bezos regularly attended O'Neill's seminars at Princeton and quickly became a devotee. Here, at last, was a way to save Earth *and* expand your profit margins: shift the whole game off the planet. If Earth is running out of resources, mine asteroids instead. If Earth is suffering from industrial pollution, move those toxic factories to the stars. And if Earth is groaning under the weight of overpopulation, send as many people as you can to live in orbital shopping malls made of whatever you find out there.

As Bezos learned from O'Neill, extraterrestrial habitats could either be constructed out of asteroidal metals or carved into the asteroids themselves. This latter possibility was the subject of a meeting of Princeton's Students for the Exploration and Development of Space (SEDS), which Bezos directed during his senior year, in the spring of 1985. As reporter Christian Davenport tells the story, Bezos spoke that afternoon to a room of about thirty people. In an energetic minilecture, he explained to his fellow students that the easiest way to turn an asteroid into a habitat was to melt its core with solar mirrors and then flood it by inserting a "massive tungsten tube" into its now-molten interior. "The water would immediately turn to steam, inflating the asteroid like a balloon—and voilà, there was your habitat."[1]

Not everyone in the room thought this idea was as thrilling as its speaker did. "As Bezos carried on," Davenport writes,

> A student in the back of the class interrupted Bezos, jumping to her feet with anger. "How dare you rape the universe!" she shouted. She then stormed out.
>
> All eyes turned to Bezos, who didn't miss a beat. "Did I hear her right?" he said. "Did she really just defend the inalienable rights of barren rocks?"

In Davenport's rendition of this story, Bezos gets the last word. This positioning leaves the reader thinking, along with Bezos and perhaps Davenport, that the student in the back was downright unhinged. What kind of person would interrupt a speaker, rather than waiting for the Q&A? How does colonizing an asteroid have anything to do with rape? And what kind of lunatic defends the rights of rocks? A rock is just a rock!

It seems so self-evident. It seems such a grammatically straightforward, fundamental, even universal truth that a rock is just a rock. After all, unlike animals and vegetables, rocks have no life, no dignity, no value—apart from whatever they might be worth on the market. Clearly, they have *that* kind of value. Clearly, rocks (along with their jazzier cousins, minerals) are important as commodities. But they're only "raw materials" until they're taken from Earth, processed, sold, and bought as building supplies, landscape enhancement, cosmetic ingredients, nutritional supplements, and fine and costume jewelry. As raw materials, rocks are ours to use as we see fit and turn into profitable products. Rocks, in short, are *resources*.

Sacred Spaces

In Australia's Northern Territory, in the sudden midst of the Outback's flat red earth, there is an absolutely gargantuan sandstone formation called Uluru. To its traditional caretakers, Uluru is sacred, having been formed by—or in some accounts, formed *of*—the ancient people, animals, and plants who created the land at the beginning of time. White explorers first laid eyes on Uluru in the late nineteenth century and named it Ayers' Rock after the then-chief secretary of South Australia. Since that time, what began as a sacred and ceremonial site for the Yankunytjatjara and Pitjantjatjara people has become a tourist destination.

When Western visitors began to flock to this stunning monolith, its Indigenous stewards asked that they treat the site with respect. They asked that visitors remain at the base of the landform, leave it as they'd found it, and refrain from taking photographs of ritually significant sites. Nevertheless, "intrepid" explorers over the last hundred years have insisted on climbing the structure, taking selfies with it, rappelling into roped-off areas, and pocketing stones to bring home as souvenirs. In 2019, the parks service finally removed the chain they'd installed to help tourists climb, but it remains a struggle to get visitors to behave themselves. They've spray-painted their names over ancient wall paintings ("SARAH!" "JADYN!"), stripped off their clothes on forbidden ground, driven golf balls from the summit, and fallen into nearly unreachable crevasses trying to get the perfect photo.

When I visited Uluru on a postcollegiate trip at the turn of the millennium, a nearby gift shop was selling bumper stickers with an image of the monolith at sunset, resplendent in its oranges, golds, and reds. The stickers were arranged in two stacks that looked identical except that one proclaimed, "I climbed Uluru!" while the other proclaimed, "I didn't climb Uluru!"

Now, I was a religion major, so there was no way I was going to climb someone's sacred mountain, especially if they were pleading with me not to. But I remember spending a long time looking at those bumper stickers with their arresting artwork and starkly opposed messages. Clearly, they were both doing what we now call "virtue signaling." Both of them wanted to let you know that the person who'd bought them had done something remarkable. But how do you weigh virtues like courage, physical fitness, and persistence ("I climbed Uluru!") against virtues like humility, respect, and concern for others ("I didn't climb Uluru!")? Was the climber's sticker also proclaiming a rugged secularity ("I refuse to be intimidated by superstition!")? And was the nonclimber's sticker perhaps commodifying its own piety ("I paid AU$4.99 to let you know how respectful I am!")?

Half a decade later, I found myself thinking about those stickers as busloads of tourists poured into the cathedral I was living in. Well, not exactly *in*, but next to—in a diocesan building on the serene grounds of New York's St. John the Divine, where I'd been appointed scholar-in-residence for the year. Every day, throngs of people would come to prowl around this infamously unfinished landmark built out of granite and limestone and said to be the world's largest neo-Gothic cathedral (at its highest point, they say, you could fit the Statue of Liberty in there). And every day, I'd see these hundreds of visitors more or less respecting the velvet ropes around the altar, more or less refraining from flash photography, and more or less speaking in the hushed tones the majestic space seemed to call for. At no point did anyone try to scale its walls. At no point did anyone pocket pieces of stained glass or try to play golf.

What is it that makes the sacredness of St. John the Divine easier for outsiders to respect than the sacredness of Uluru? Part of it is familiarity: most of the Cathedral's visitors are at least minimally aware of Christian traditions, whereas most of Uluṟu's visitors are completely ignorant of Aboriginal traditions. Part of it is bald-faced racism: the structures reminiscent of the Roman and British empires command more respect among wealthy globetrotters than the structures stewarded by nonimperial peoples. And beneath it all, pervading the Western unconscious, is what you might call an *antimineralism*: a tendency to value those rocks that have been removed, installed, carved, stacked, and shaped by human hands (and market forces) over those rocks that remain where and as geological (and ancestral) processes made them.

Like every unconscious preference, this antimineralism is the product of a history that's often hard to see. Chapter 2 explored this history through the work of Lynn White, who linked the escalating climate disaster to the rise of modern technology and the rise of modern technology to the Christian "victory over paganism." According to the Indigenous worldviews that imperial Christianity continues to denigrate as "pagan," rivers might be persons, forests might be spirits, mountains might be ancestors, and stones might dance and talk. I say "might" because different communities value different natural forms differently. A stone that is sacred to some people is not sacred to others; a fish that is kin to some people is a food source for others; an herb that's taboo for one culture is either unknown or unrestricted in another; Earth, a goddess for some, is an ancestor or sibling for others; and so forth. The point is that all of these diversities vanish with the Christian victory over paganism, which unilaterally declares animals, plants, and especially minerals to be only important as raw materials for the human conquest of a lifeless Earth.

This is what White means when he says that Western technology relies not only on Western economics but also on Western *religion*: if rocks and rivers and trees were alive, if nonhuman animals were persons, then it would be impossible to exploit them for maximum production. You can't strip-mine a mountain you affirm as holy no matter how much quartz might be in there. You can't dam and divert a river you talk with or pray to no matter how much cheap energy it might give you. And a sacred forest will let you use its wood, but only if you ask nicely, take only what you need, and contribute in return to its regeneration. So the only way you can take everything you want from Earth is by insisting Earth has no wants of its own. That Earth isn't a goddess, an ancestor, or even a living system, but just an inert container of inert objects waiting to be extracted, processed, and turned into profit.

At this point, we can appreciate the overlapping collusions of imperial Christianity with technology, capitalism, and colonialism. As the alleged crown of creation, (European) humans believed they had not only the right but the duty to dominate Earth, to expand Christendom into Africa, India, Australia, and the Americas while making full use of these continents' human and nonhuman "resources." Anyone who might object on behalf of Earth and its people was a heretic, idolater, Canaanite, or pagan destined for conversion, elimination, or both.

These days the people who object on behalf of Earth often include Christians themselves who have joined the water protectors at Standing Rock, protested fracking in southern England, opposed the Line 3 pipeline in Canada and Minnesota, formed local chapters of Extinction Rebellion, converted church greens into meadows, and established organic farms and animal sanctuaries in upstate New York. Such Earth protectors include members of most "mainline" denominations (the Lutherans, Episcopalians, Methodists, etc.), a good portion of younger Evangelicals, and the current pope, who condemns the notion that Earth's creatures are "mere objects subjected to arbitrary human domination . . . a source of profit and gain."[2] So the Christians, for the most part, are coming around.

Meanwhile, the people toeing imperial Christendom's "human dominion" line are atheist billionaires who insist that their investors' bottom lines are more important than wetlands, barrier reefs, drinkable water, and breathable air—let alone mountains or asteroids, which only a pagan hysteric could love.

Did I hear her right? Did she really just defend the inalienable rights of barren rocks?

I don't know the woman who stormed out on Bezos that afternoon at Princeton, but I find myself thinking about her every time another space-mining company makes the news with its promises of infinite resources. *How dare you rape the universe!*

I wouldn't have put it so boldly, I fear, or so memorably. But in my braver moments, I think I can feel a bit of that blinding rage. It's the rage of someone who's sick over the decimation of this planet and horrified that this planet isn't enough for the decimators. It's the rage of a space geek (she attended that meeting, after all) who's perhaps seen in the Moon and the stars another, more peaceful way of being and who can't bear the thought of exporting the tungsten tubes, mine shafts, and oil drills that penetrate our ravaged "Mother Earth" out to the unsullied solar system. It's the rage of a woman trying to make it in the almost all-male sciences in the mid-1980s at a historically white university on stolen Lenni-Lenape land in a country that can't even pass a damn Equal Rights Amendment. *And as if all that shit wasn't enough*, we might shout with Gil Scott-Heron, this rat now wants the asteroid belt. With Whitey on the Moon.

Cosmic Gold Rush

The plan first took shape in the mind of the omnipresent Wernher von Braun. The recent German expat and more recently born-again Christian argued that, to preserve its military leadership, boost its economy, and fulfill its divine calling as a chosen nation, the US would need to expand into outer space. "I see space as the endless frontier," von Braun wrote in a *Popular Science* essay, "more exciting than the discovery of the New World or the conquest of the American West."[3] His plan, now known as the "Von Braun Paradigm," would unfold in four successive stages: first, we'd build a reusable space shuttle, then an orbital space station, then a permanent base on the Moon, and finally we'd be able to establish a human presence on Mars. (If these stages sound familiar, it's because—however historians may quibble over details—the Von Braun Paradigm has guided US space policy since its progenitor wrote those *Colliers* essays in the early 1950s.[4])

Crucially, it was von Braun who argued that the key to taming the "endless frontier" of space would be capturing its endless "resources." As he promised just a few years after Apollo 17 returned with the missions' last lunar samples, "minerals can be mined on the moon, in the asteroid belt, and on nearby planets in the next century."[5] Such minerals could eventually be used in situ, he suggested, to build "permanent inhabited bases on the moon or Mars, and orbiting colonies of hundreds of thousands of people." And just like the earthly frontier, this cosmic frontier would give its pioneers and pilgrims "new places to live—a chance to organize a new interplanetary society, and make fresh beginnings."

To anyone familiar with Western imperial history, however, it's clear that the frontier's "fresh beginnings" always begin at someone else's expense. The "discovery of the New World" and "conquest of the American West" relied on the forced labor of enslaved people. So did von Braun's V-2 rocket facility in northern Germany's Pennemünde, which ran on the strength of the sixty thousand workers von Braun helped select from concentration camps, twenty thousand of whom died on the job from "assault, starvation, and sickness."[6] The "fresh beginnings" of the New World and the American West took place, moreover, on land stolen from its traditional stewards, whom the self-appointed people of God saw as Canaanites on the wrong side of history.

So, as the environmentalist poet Wendell Berry fumed when he heard about O'Neill's space habitats hewn from the metals in captured asteroids, there is nothing particularly "fresh" about the effort to colonize space or about the kinds of lives it will enable. "What is remarkable about Mr. O'Neill's project," Berry wrote, "is not its novelty or its adventurousness, but its conventionality."[7] Behind this vision of conquest, Berry detected the old European American "idea of Progress, with all its old lust for unrestrained expansion . . .

its exclusive reliance on technical and economic criteria . . . its compulsive salesmanship."

Compulsive. As if there's no other way to behave, no possible end to the frenetic rituals of conquering, extracting, buying, and selling. As if something terrible will happen if we stop "progressing" toward the same amorphous, endless end of Progress.

As we've seen in Bezos and Musk alike, this "something terrible" is the end of what we've come to call "the American way of life." It's the end of endless cheap energy and cheaper clothes and gadgets, the end of endless consumption and profit. Unable—or maybe just unwilling—to imagine any other way to flourish, companies like Moon Express, iSpace, Lunar Outpost, Deep Space Industries, and Planetary Resources have sprung up to answer the cosmic call of capitalist duty. In collaboration with the European Space Agency—as well as the national space agencies of Japan, China, Russia, Luxembourg, the UAE, and the US—these off-world speculators are promising what physicist Michio Kaku has called "another gold rush in outer space."[8]

At the moment, it's not quite clear how zero-gravity gold mining would actually work. NASA's Robotic Asteroid Prospector (RAP) project is studying the possibility of intercepting or redirecting asteroids—not only to prevent them from hitting Earth but also to send them into orbit around the Moon so they're easier for miners to access. That having been said, it's still not clear whether asteroids are even solid enough for large machinery to land on or whether they might break apart under too much force. So the first step is to figure out how to interact with them.

In 2016, NASA launched a billion-dollar probe it called OSIRIS-REx, its name a strange combination of the Egyptian god of rebirth and the Latin word for "king." Two years later, OSIRIS fell into orbit around Bennu, a carbonaceous asteroid about the size of the Empire State Building that's closer to Earth than most of its cousins. Like the probe designed to collide with it, Bennu is also named after an Egyptian god: a birdlike deity who hovered over the primordial waters and called for the world to come into being. Said to be the precursor to the Greek phoenix and a symbol of Osiris, Bennu is associated, like Osiris, with rebirth—with life that comes out of the ashes of death. What is NASA trying to tell us with these names?

Well, a lot of things. As NASA explains it, "Bennu" was suggested by a nine-year-old North Carolinian named Michael Puzio, who won the agency's cosponsored *Name That Asteroid!* competition.[9] To Puzio, the solar panels and sampling arm of the OSIRIS probe looked like Osiris's symbol, Bennu. Moreover, as NASA's website elaborates, Bennu's role in creation is particularly salient to the scientific heart of the mission, since "asteroids may harbor hints about the origin of all life on Earth!" Such "hints" may be embedded in the primordial mixture that this (until now) undisturbed orbiting fossil preserves: the carbon, oxygen, hydrogen, and nitrogen that somehow gave way to terrestrial microbes 4.5 billion years ago.

And then there's the other stuff the asteroid contains. There's the organic mystery, the website tells us, "But also platinum and gold!" And it's here that we get to the heart of the resurrection Bennu and Osiris are promising. "In lieu of Earth's finite resources," which clearly can't sustain us forever, NASA explains that "many asteroids do contain elements that could be used industrially" to support our rebirth as cosmic citizens. And while they joke about creating "extraterrestrial jewelry" out of the stuff they mine from asteroids, NASA explains that the most important resource in Bennu is water (a cross-cultural symbol of rebirth), which can be used to keep astronauts alive on long-haul journeys and to create rocket fuel to send humans farther into space.

When it comes to asteroids, NASA tends to retain its jovial, pragmatic tone: sure, they contain enough gold to make Madison Avenue blush, but our real concern is learning about our origins on Earth and our future in space. News media and popular science writers, on the other hand—along with senators and high-risk investors—tell us we're on the verge of something economically huge. As Kaku explains it, "asteroids . . . are like flying gold mines in space."[10] Bennu may be a carbon-heavy "rubble pile," but its neighbor Psyche seems to be solid metal, and 2011-UW158 seems to contain $5 trillion worth of platinum.[11] If the surveyors' calculations are right, then your average asteroid should contain, in addition to all that water, multiple billions of dollars' worth of industrial metals like copper, nickel, and lead; precious metals like platinum, silver, and gold; and to top it all off, rare-earth metals like lanthanum, dysprosium, and gadolinium.

Mostly discovered during the nineteenth century, rare-earth metals are used in very small amounts to enhance the performance of all sorts of technologies. Known as the "spices" or "vitamins" of modern industry,[12] these rare earths have become indispensable to the production of lasers, smartphones, medical imaging equipment, plasma screens, fuel cells, refined petroleum, stress-resistant crops, wind turbines, and electric cars. As such, rare earths are often the rhetorical linchpin for space-mining advocates, who need only mention their name to convey the ominous scarcity that compels us to forage for space treasure.

But here's the thing: rare earths aren't rare. They're not renewable, of course, but they could be recyclable and even reusable if we put our minds to it, and most of them are abundant in Earth's crust. Their name is a remnant of the long, confusing process of their discovery, and at this point it's become a misleading one. As geographer Julie Michelle Klinger maintains, "we are nowhere near exhausting Earth of potential mining sites, for rare earths or otherwise."[13] So why does everyone keep saying that we are?

The story is a remarkably recent one. Since their discovery, rare earths have been mined in places as mutually far-flung as India, Brazil, South Africa, California, Malaysia, and Australia. Since these elements tend to be found in deposits that also contain heavy metals like arsenic, uranium, and fluoride, however, they are difficult and dangerous to extract. Lax environmental regulations— who wants to stand in the way of such a lucrative industry?—have allowed mining companies to sidestep the pollution-control measures that might protect the water and soil they contaminate, causing "cancers, birth defects, and the decomposition of living people's musculoskeletal systems."[14] By the 1990s, enough costly environmental disasters had piled up around the globe that rare-earth manufacturing moved almost entirely to China, which ramped up production at its Bayan Obo mine in Inner Mongolia's Baotou Region. By 2010, China was supplying 97 percent of the world's supply of rare-earth materials.

For decades, the West was unperturbed by China's concentrating the toxic burden of rare-earth mining on ethnic Mongolians. As Klinger tells the story, however, everything changed in September of 2010, when China's military blocked an otherwise unremarkable shipment of rare-earth cargo headed to Japan and created enough of a backlog to send waves of panic through the rest of the industrialized world. Prices soared—in the infamous case of dysprosium, by 2,000 percent. American politicians and news outlets suddenly decried the nation's "dependence on China" as a "national security threat" and demanded that industry do something to loosen the "stranglehold" Beijing held on the rare-earth economy.[15]

Enter the space-mining companies, whose CEOs finally found a justification for their literally outlandish business models. Sure, they admitted, it'll be hard to mine asteroids. But it will be worth it for the freedom from China, the freedom from environmental protocols, and, you know, for the trillion-dollar asteroids. Granted, the rocks themselves are too heavy to bring all the way back to Earth (we couldn't even bring Buzz Aldrin's diapers), so the firms propose building in situ refineries to extract the valuable stuff from the rocks we don't need. Sounding a lot like young Bezos at Princeton, the engineers explain that all you have to do is inject hot fluid into the ancient boulders, melt their metals, extract the desirable solute from the instrumental solvent, and throw the rest away. ("Where will they throw it away," you ask? Who knows! *There's plenty of space out in space.*)

For any of this to happen, of course, we're going to need infrastructure. You can't process metals without facilities for doing so, which can't be built without metals, so until in situ resource utilization is up and running, we'll need to build outerspatial factories with materials from Earth. You can't get the materials "up" to the

Moon or out to an asteroid unless private actors keep prices down by competing for contracts. You can't get private actors to compete for *anything* without guaranteeing a return on their investments. And you can't guarantee a return on investments if international law seems to say you're not allowed to own anything in space.

So it all comes together like some diabolical 4-D puzzle: nations like the US, UAE, and Luxembourg go rogue, making their own legislation to protect the investors who fund the private companies that drive down prices so the space agencies can hire them to mine the materials to build the colonies that send more probes prospecting for space gold on the endless frontier, with all of it backed up by a dedicated branch of the military.

Terraformal Dreaming

As we've already seen, the question of what this whole space adventure is *for* turns in on itself. The best NASA can do is to say that we're gathering the resources to go to space so that space can provide the resources we need to live in space. And before you have time to realize it's a circular argument, they tell you we're doing all of this mining and mooning so we can one day get to Mars.

Mars! Not the most hospitable planet in the galaxy. (That seems to be Earth.) First of all, Mars is freezing: the average temperature is around 80 degrees below zero on the Fahrenheit scale (−62°C), a little colder than Antarctica. In the summertime, some places can reach 70°F (21°C), but in the winter, temperatures at the poles can drop to −220°F (−140°C), which is significantly colder than anything humans have recorded on Earth. And the temperature is just the beginning.

The atmosphere on Mars is 95 percent carbon dioxide, so we couldn't breathe there. The atmospheric pressure is too low to support any sort of life we've learned about on Earth. The soil is superfine, toxic to plants and animals, and prone to forming "dust devils" that whirl across the surface at 30–60 miles per hour, leaving an atmospheric haze in their wake for days or even weeks. And to top it all off, there's no magnetic sphere, so the Red Planet is constantly bombarded by carcinogenic solar and galactic radiation.

For anyone who's wondering how bad Mars could actually be, science writer Ross Andersen explains, "if you were to stroll onto its surface without a spacesuit, your eyes and skin would peel away like sheets of burning paper, and your blood would turn to steam, killing you within 30 seconds." (*Saturday Night Live* viewers might recall the stomach-churning scene from the episode Elon Musk hosted in May of 2021, when Pete Davidson's Chad tries to remove his helmet and his face explodes.) "Even in a suit," Andersen continues, "you'd be vulnerable to cosmic radiation, and dust storms that occasionally coat the entire Martian globe in clouds of skin-burning particulates, small enough to penetrate the tightest of seams."[16] In short, as SpaceX president Gwynne Shotwell admits, Mars is a bit of a "fixer-upper of a planet."[17]

Okay, but how do you fix up a planet? Isn't this already the question on everyone's mind when it comes to our poisoned Earth? How do you scrub the CO_2 out of the atmosphere, the acid out of the rain, the plastic out of the oceans, the infestations out of the trees, those few degrees out of the climate? Well, it turns out that the same processes currently destroying life on Earth might actually create it on Mars. At least, that's what the Red Planetarians tell us. As we saw in the introduction, Robert Zubrin says that the trick to making Mars habitable would be "intentionally manufacturing a greenhouse effect" akin to the one we've unintentionally manufactured here. It would just be a matter of "producing fluorocarbon super-greenhouse gases on Mars . . . and willfully dumping these climate-altering substances into the atmosphere."[18]

What could possibly go wrong?

Zubrin's plan is one instance of a process called *terraforming*: the hypothetical recreation of a planet in the image of Earth. At first, humans on Mars would need to live in underground bunkers, grow food in inflatable greenhouses, figure out how to manufacture things in 37 percent gravity, and send rovers to find water and metal deposits. Once they'd harvested enough in situ materials and learned how to make bricks, plastics, glasses, metals, and ceramics, the Martian pioneers could build pressurized structures above ground and move into "domains the size of shopping malls."[19] In the meantime, they'd be working to raise the planet's temperature to a balmy 32°F (0°C), probably about as much as you can ask of this fourth planet from the Sun.[20]

Proposals for warming up Mars vary dramatically. Some people join Zubrin in suggesting that tons of chlorofluorocarbons unleashed throughout the globe might just do the trick. Others focus on the frozen ice caps, which contain both water and carbon dioxide. We could melt the caps with orbital mirrors, or paint them with black soot, or cover them with lichens, or blast them with solar-powered space lasers . . . or, if Musk has his way, "nuke" them with ten thousand missiles. Either way, planetary scientist Christopher McKay estimates it would take about 100 years to make the planet as warm as an ordinary springtime in Alaska.[21]

But that's just the temperature. Air is another matter entirely. Creating a breathable atmosphere on Mars would require far more than just heat; in fact, it would probably require the same mixture of hydrogen, carbon, oxygen, nitrogen, phosphorous, and sulfur, along with "methane, ammonia, formaldehyde, sulfides, nitriles, and simple sugars," that gave rise to life on our own planet.[22] As evolutionary biologist Lynn Margulis demonstrated along with Earth scientist James Lovelock, these crucial substances are regulated by protobiotic life-forms themselves.[23] That means that if we want to bring Mars to life, we're going to need a planet's worth of microbes, whose complex interactions will produce the conditions life needs to survive, flourish, and evolve. And these microcreators would take anywhere between ten thousand and one hundred thousand years to shape Mars into the kind of planet across which big animals could walk or swim or hoverboard without a space suit.[24] So it's going to be a long time living in those shopping malls.

The question, then, is, Should we do it? (Assuming we can?)

According to Zubrin, we not only should, we *must*. Rooted as he is in the long history of American conquest, Zubrin believes that "the creation of a new frontier [is] America's and humanity's greatest social need. Nothing," he writes, "is more important."[25] Zubrin gets the idea from the early twentieth-century historian Frederick Jackson Turner, who argued that the identity of the US has been fundamentally shaped by its progressive frontiers. According to Turner, America gained a sense of itself apart from its European ancestors as it expanded from the Eastern Seaboard across the vast Midwest, up to the Great Lakes, down to the Mississippi delta and all the way out to the Pacific ("each," he boasts, "was won by a series of Indian wars").[26] For Turner, each victory of the new nation over its European and Indigenous rivals contributed to America's "perennial rebirth," increasing the national character in strength, independence, energy, and freedom. By the end of the nineteenth century, the conquest was complete, and the question for Turner in 1893 was how America could retain its character now that "the frontier has gone."

For Zubrin, the answer is easy: it can't. The signs, he insists, are everywhere: economic decline, technological stagnancy, bureaucracy and overregulation, reality television, loss of overall vigor: it's clear the US is a nation in decline. We've got to get back to our roots, Zubrin cries, and our roots are on the open frontier. "Without a frontier from which to breathe life," he writes in a tribute to Turner, "the spirit that gave rise to the progressive humanistic culture that America has offered to the world for the past several centuries is fading."[27]

In line with Johnson, JFK, Trump, and Pence alike, we might notice that Zubrin here is equating American flourishing with human flourishing. The US has offered the world a model of peace, decency, human rights, and freedom, but it's only the frontier that's allowed the US to model those characteristics in the first place—or so the reasoning goes. "Without a frontier to grow in," Zubrin insists, "not only American society, but the entire global civilization based upon Western enlightenment values of humanism, reason, science and progress will die." To his mind, the fate of any "humanity" worth mentioning rests on the fate of America, the fate of America rests on the opening of a new frontier, and "humanity's new frontier can only be Mars."

"Mars?" asks the *Atlantic*'s Shannon Stirone in an article I've probably read fifteen times; "Mars is a hellhole."[28] Responding to Elon Musk's lust for immortality on the Red Planet, which is largely inspired by his mentor Zubrin, Stirone cites the usual list of unsavory Martian attributes: the lack of pressure, the freezing temperatures, the seeming lifelessness, the blood-boiling atmosphere. Mars, she argues, isn't going to save us. "Mars," she counters, "will kill you."

For Zubrin, however, these atmospheric challenges are just what the languishing American spirit needs to reinvigorate itself. Every site on Earth is too easy, too decadent, not to mention too regulated to serve as a genuine frontier. (Even in Antarctica, he says, "the cops are too close.") Venus is too damn hot. The Moon doesn't have enough of the elements we need to survive. Mars, on the other hand, "possesses oceans of water frozen into its soil as permafrost, as well as vast quantities of carbon, nitrogen, hydrogen, and oxygen, all in forms readily accessible to those clever enough to use them."[29] And if we don't act soon—here comes the urgency—we'll lose the collective cleverness we need to figure it out. "Mars today waits for the children of the old frontier," he says. But considering how stupid Americans are getting, how soft we're growing around our pioneer edges, "Mars will not wait forever."[30]

So we must act now. "Failure to terraform Mars constitutes a failure to live up to our human nature," Zubrin writes. Should we succeed, however, we will not only fulfill this nature; we will surpass it. In the process of creating life on a dead planet, the (American-led) human species will approach divinity, making worlds out of nothing at all. "Gods we'll never be," says Zubrin, in the midst of the suddenly biblical flourish that culminates his *Case for Mars*, but we can become "more than just animals."[31] Terraforming the Red Planet will reveal humanity to be creatures who "carry a unique spark," who are godly *enough* to provide a new planetary home for "the fish of the sea . . . the fowl of the air, and every living thing that moveth upon the Earth."

Moveth! At the culmination of his "let's terraform Mars" book, Zubrin cites not just the Bible but the King James Version, as if the arcane language might remind "humanity" of its God-given governance over "the fish of the sea" and every other creature. As if it might rekindle the "unique spark" of Genesis 1, which made Man in the image of God and now promises to make worlds in the image of Man. In case we've missed the point, Zubrin makes all these biblical references more explicit in a later article, writing that terraforming Mars would constitute "the most profound vindication of the *divine nature of the human spirit*—dominion over nature, exercised in highest form to bring dead worlds to life."[32] God's Newest Israelites will be cosmic necromancers.

Younger advocates of terraforming, all of them influenced by Zubrin, tend to focus less on the power of creation than salvation. For these newer terraformers, Musk among them, the imminent danger we face is not just the decline of Western civilization but the death of the human species itself. If some cataclysmic event were to wipe out life on Earth, they reason, everything worth anything would be gone. Therefore, it's important to plant as many cosmic colonies as possible, to increase the likelihood that one of them might survive. This is what Musk means when he professes his "duty to maintain the light of consciousness, to make sure it continues into the future."[33]

Other advocates of terraforming go even further than Musk, expressing their intention to save not just humanity but *life itself* by giving it another place to take root. After all, when the Sun explodes in five and a half billion years, the game will be over for anything that eats, breathes, or excretes. And sure, a dying Sun will do away with Mars as well as Earth. But Mars can serve as an eventual launching pad to another solar system. Therefore, as geologist Martyn Fogg concludes, "total extinction of terrestrial life can . . . only ultimately be avoided by vacating our planet for a more benevolent locale elsewhere in the cosmos."[34]

A more benevolent locale? Seriously, what are the chances? We don't know a ton about the planets that orbit other stars, but the data so far isn't all that promising. Some exoplanets seem to reach 1700°F (927°C). Some are losing their atmospheres to overactive solar radiation. Many of them orbit red dwarf stars, frigid little things that periodically fire forth solar flares to obliterate any life that might be trying to live. So, as Carl Sagan warned in the eco-cidal 1990s, the "pale blue dot" we were born on—with its oxygen, nitrogen, liquid oceans, and plentiful sunshine; its forests and waterfalls and calling birds and fragile bees—seems still to be the most benevolent planet out there.

Perhaps we could resolve to terraform Earth, instead? In the words of Lucianne Walkowicz, "If we truly believe in our ability to bend the hostile environments of Mars for human habitation, then we should be able to surmount the far easier task of preserving the habitability of the Earth."[35] Why not put all that money, energy, and manly frontierism into bringing our own ecosystem back to life?

Faced with this sort of argument, terraformers who claim to love their home planet (Musk is not among them) promise that hacking Mars will ultimately be good for Earth as well. With smart-sounding proposals like "comparative planetology," they claim that figuring out how to bring life to another planet will also help us save life on this one.[36] But this argument perplexes me. We already know what we have to do to save life on this planet. We have to reduce carbon emissions, stop using plastics, plant more trees, clean up the oceans, restore the rainforests, ban industrial farming, eat as little meat as possible, and divert subsidies away from oil and cars and toward public transportation. We don't need folks to aspirate on Martian dust and die of radiation poisoning in order to learn any of that. The terraformers just don't seem to want to know what everybody knows, which is that the extraction, consumption, and runaway pollution that are wrecking the life on this planet won't suddenly become lifesavers when we export them to Mars.

Suspicious of the techno-tactics of terraforming, some of the more ecologically sensitive spaceniks suggest an alternative in *eco-*

poiesis. Literally meaning "home building," initiating ecopoiesis would mean sending a flood of microbes and chemicals to do the long work of allowing more complex life to evolve (or not) on Mars. Far from turning the planet into the terraformers' shopping mall, garden paradise, or cosmic amusement park, Margulis explains that ecopoiesis "would transform [Mars] into a global cesspool—colorful, perhaps, but rich in mephitic vapors" and smelling like "sewer gas."[37] Such a protobiotic planet would not in any sense be *for* humanity—at least not for tens of thousands of years—but it might well be in service of some sort of life. So perhaps instead of sending rovers and drills and refineries, we should send the primordial "goo" and see what happens?[38]

As far as Carl Sagan was concerned, it all depends on what's on Mars in the first place. Most arguments in favor of terraforming and ecopoiesis assume that, although life may have existed on Mars in the past, it no longer does. But it's always possible that microscopic creatures are hiding somewhere under the ground or inside the rocks. And in that case, Sagan wrote, "if there is life on Mars, I believe we should do nothing with Mars. Mars then belongs to the Martians, even if the Martians are only microbes."[39] As astrobiologist David Grinspoon has argued, however, our understanding of "life" is based exclusively on terrestrial biota. We might not know how to recognize life on Mars even if we looked straight at it.[40] So at what point would we decide Mars is sufficiently "dead" to ecopoieticize it or terraform it? How would we ever know we weren't interfering with the native biotic processes of Mars?

For Zubrin and Musk, it doesn't matter. Whatever microbial life might exist in a frozen rock on the Red Planet, it's nothing compared to the complexity of terrestrial vegetation, animal diversity, and human arts on this blue-green one. So we have an obligation to spread our earthly kind of existence, no matter what sort of existence Mars might be trying to sputter out.

At this point, a whole new coalition of ethicists begins to protest, telling us to leave the rocks alone.

Cosmic Vandalism

As we've seen, Wernher von Braun knew as soon as those Moon rocks returned that extraterrestrial minerals would build our future in space. And although investments in cosmic mining didn't really get going until that China-led shipping crisis in 2010, the economic writing has long been on the outerspatial walls.

Six months after her first flight aboard *Challenger*, Sally Ride told Gloria Steinem in an interview with *Ms.* Magazine that NASA would never get the funding it needed without the help of private industry. But, Ride predicted, it was only a matter of time. At the behest of two big pharmaceutical companies, Ride and her colleagues had conducted a few experiments to confirm that some medications can only be made in a weightless environment. So, she told Steinem, "as soon as people start to realize that they can be making a lot of money by taking advantage of [space], then I think that private industry will probably take over and help with the expansion [of the space program]."[41]

Perplexingly, the ever-pugnacious Democratic Socialist Steinem said nothing in response. Maybe she was genuinely starstruck by the encounter with America's first woman astronaut? Or maybe the idea just seemed too far-fetched to worry about, not even worth contesting. How could the national space administration of the United States of America ever defer to big pharma and a bunch of cocksure start-ups chasing moon rocks? And yet here we are, with NASA buying lunar dust from corporations so it can settle the Moon, mine those asteroids, and get to Mars with military backup for the sake of the burgeoning space economy.

To its opponents, the problem with the escalating gold rush in outer space is that it presumes the whole universe is ours for the taking. As space theorist Natalie Treviño points out, for all its talk of "common heritage" and "peaceful purposes," even the Outer Space Treaty "binds celestial bodies to exploitation."[42] By declaring everything out there to be "the province of all mankind," the OST's highest aim is to allow every nation to "use" outer space without fighting and to benefit equally from this exploitation. In the minds of the opponents of space mining, then, we've got it all backward. As the philosopher Holmes Rolston III puts it, we keep asking how to make outer space into "resources," which is to say, how "this astronomical world can *belong to us*," when what we should be asking is "how we *belong to it*," and "whether it *belongs to itself*."[43]

How we belong to it. How we might live thoughtfully into our tiny place in the infinite cosmos. How we might ask what it needs from us rather than what we want from it.

Whether it belongs to itself. Whether asteroids might prefer that we not core them out for profit. Whether Mars and the Moon might not, in fact, be up for grabs. Whether rocks themselves have rights.

I know it might sound a bit much. Earthlings are still struggling to secure the basic rights of migrants, Black and Indigenous people, children, women, queer and trans folk, and workers—to name a few. Millions of animals on the planet don't have bodily autonomy or the right to live out their days in peace. And here I am, asking you to consider the rights of rocks. So let me say from the outset that I don't think it's ethically *necessary* to say that rocks have rights in order to envision a just and peaceful approach to outer space. But I do think we can learn a lot from asking whether it's possible. What might it mean to say that a rock "belongs to itself"?

The laws, ethics, technologies, economies, religions, and languages of the West tend to divide the world into subjects and objects. Subjects are creatures who *act*: they move, make things, destroy things, express desires, and attempt to fulfill them. Objects are the things that are *acted on*: they get moved, made, destroyed, desired, and deployed. When social movements try to secure rights for a traditionally exploited community, their strategies usually revolve around the subjectivity of objectified bodies.

To name just a few examples, eighteenth century abolitionists insisted that enslaved people were humans rather than property; the early feminist movement railed against "the objectification of women"; and contemporary PETA ads feature animals saying, "I'm ME, not MEAT." Since objects are exploitable by definition, the hope in all these cases is to widen the category of "subject." "Ham" is a product you process, sell, buy, and consume. "A pig" is a being who likes to be clean, wags her tail when she's happy, grunts little songs to her piglets while she's nursing, and as such deserves not to be penned up in squalid conditions only to be sent to a slaughter-house. In all these cases, the pursuit of justice seeks to turn objects into subjects.

For Rolston, however, it's not always helpful to divide the world into these two categories, especially if it seems to give subjects permission to exploit whatever they see as objects. When it comes to the more-than-human world, Rolston suggests we understand nature neither as subjective nor objective but as *projective*, meaning that it makes *projects*. These projects include not only animals, vegetables, and microbes but also lakes and rivers, moons and stars, planets, comets, and asteroids. As projects, these natural entities have their own value independently of anything humans might want from them.

Does this mean we can't eat a carrot, dig up a weed, or burn a lump of charcoal, since all of them are projects of nature? No. But it does mean we ought to spend some time asking what we're allowed to take from the universe. For his part, Rolston offers six criteria for "respectful" treatment, by which he seems to mean that we shouldn't destroy, deface, poison, deplete, or abuse the project in question.

1. We should respect any natural place that we've named. As Rolston explains, if we've given a proper name to a moun-tain, crater, hill, or valley, then we have already acknowl-edged it has a "topographic integrity" that makes it worth preserving.[44] This doesn't mean we can't ever alter anything we've named. It just means that protecting the forma-tion ought to be at the very least part of "the calculus of tradeoffs" involved in our decision to proceed. If a public or

private entity is trying to figure out whether or not to mine a Martian mountain (we've named all of them), someone should be thinking not only of the potential financial risk, the potential financial benefit, and the well-being of the workers but also of the well-being of the mountain itself.

2. We should respect "extremes in natural projects." The Valles Marineris on Mars is "four times as deep as the Grand Canyon and as long as the United States is wide."[45] We probably shouldn't turn it into a parking lot, a garbage pit, or the galaxy's biggest swimming pool.

3. We should "respect places of historical value," by which Rolston means not just human history but also geological history. He gives the example of Callisto, Jupiter's second-largest Moon, whose extraordinarily low temperature makes it an effective "ice museum" containing clues to the origins of the solar system. Again, "respecting" Callisto doesn't mean staying away from it at all costs. But it certainly means refraining from melting all its ice into rocket fuel and then moving on to the next cosmic gold mine.

4. We should "respect places of active and potential creativity." If it looks like more biotic and nonbiotic "projects" might be underway within a cosmic body, we should let them unfold without disturbance.

5. We should "respect places of aesthetic value." Any natural formation that causes us to gasp, cry, marvel, dance and sing, or photograph it until we run out of batteries should probably be preserved.

6. We should "respect places of transformative value," by which Rolston means places that change the way we look at the world or understand ourselves.

As both critics and followers of Rolston point out, terraforming Mars would violate all six of these principles.[46] We've already named every large landform and valley on the planet; many of Mars's formations are "extreme" by earthly measures; even its smallest rocks contain clues about its past (which may have included a thriving biosphere); these rocks might contain Martian microbes; the landscape is by all measures breathtaking; and anyone who encountered it would almost certainly be existentially, spiritually, intellectually, and physiologically transformed.

For these reasons, philosopher Robert Sparrow states that terraforming Mars would make human beings into "cosmic vandals."[47] By turning the whole planet into resources for human consumption, we would be refusing to acknowledge the beauty and integrity of Mars itself. We would, in fact, be demolishing it in an attempt to recreate it. Terraformation would in this sense be an egregious act of "hubris," that sin of trying to be like the gods. And as Sparrow reminds us, hubris "is traditionally punished by disaster." Floods, sulfuric fires, and untimely death tend to befall those who try to rival the gods.

Especially given our terrible track record with planetary management on Earth, Sparrow reasons that any attempt to terraform Mars would almost certainly go wrong, producing a planet "with a poisonous atmosphere or without water or lashed by continual typhoons." Thanks, in fact, to the second law of thermodynamics, it is much easier to produce chaos than it is to create order. So in the case of a whole biosphere, which would require an exquisite balance of chemicals, forces, and compounds, "anything other than complete success would be a disaster."[48]

One particularly instructive example is that of Biosphere 2. Built from 1984 to 1991 in Arizona's Sonoran Desert, this sprawling complex occupies a seventeen-acre footprint, 3.15 acres of which are indoors. By the end of September 1991, Biosphere 2 also contained enough soil, air, water, plants, animals, and supplies to sustain eight human beings, who intended to grow their own food, produce their own oxygen, and live inside the sealed glass-and-steel structure for two full years supported by nothing from the outside except a steady power source.

Infamously, the experiment didn't work. Not only did one of the eight "biosphereans" need to leave a few weeks into the experiment for medical attention (she returned a few days later, incurring criticisms that the project had already failed), but it gradually became impossible to produce enough food, water, and air to sustain eight people in a closed system. As one postmortem paper explains, "1.4 years after [the] material closure of Biosphere 2, the oxygen concentration . . . fell from 21% to about 14%." The crew broke their enclosure again to import extra oxygen, but carbon dioxide levels continued to climb out of control. In the meantime, all the large trees were overtaken by parasitic vines and "all pollinators went extinct," along with "the majority of the introduced insects . . . leaving crazy ants running everywhere, together with scattered cockroaches and katydids."[49] If this experiment in terraforming failed despite its having started out with a perfectly breathable atmosphere, a full range of living creatures, and an unlimited external supply of energy, how do we imagine things will go on a planet with none of these things, six or eight months away from anyone who might help?

Here I've tumbled into what philosophers call a functionalist argument, suggesting that we shouldn't try to terraform Mars because Biosphere 2 has taught us that it probably won't work. But the question at hand is not whether or not we can terraform Mars. The question is, *even if we could* create a perfect Martian replica of Earth, should we? Or would doing so make us "cosmic vandals"? Would the Red Planet be better off in human hands, or is it important to keep Mars Martian?

The only way I know to answer this question is to intensify it—by bringing it closer to home.

Everyone's Gone to the Moon

From space tourism to asteroid mining to ecopoiesis and terraforming, all of NewSpace's adventures depend first on settling the Moon. NASA plans to have a permanent outpost there by 2028, China has landed a rover on the Moon's far side with its own mini-biosphere experiment (including cotton, rapeseed, potatoes, rock

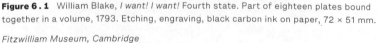

Figure 6.1 William Blake, *I want! I want!* Fourth state. Part of eighteen plates bound together in a volume, 1793. Etching, engraving, black carbon ink on paper, 72 × 51 mm.

Fitzwilliam Museum, Cambridge

cress, yeast, fruit flies, and silkworms), China plans to finish its own lunar base by 2030, Russia wants a full-fledged colony by 2030, Japan is planning a settlement it calls Moon Valley, India is refining its lunar landers, and the European and Chinese space agencies are teaming up to create a Moon Village sometime in the 2020s (a collaboration in which the US can't take any part thanks to the 2011 "Wolf Amendment" that prohibits NASA from spending any of its money to work with China).[50] And all of these national space agencies are relying on the assurances of corporations like iSpace, which promises both public and private actors "access [to] new business opportunities on the Moon," and Moon Express, whose

CEO pledged to the US House of Representatives that he'd turn the Moon into "a gas station in the sky."[51]

Just as a reminder, the plan as outlined in the vision of NASA's Artemis program is to find and extract the Moon's water deposits, convert the water's oxygen and hydrogen into rocket fuel, use whatever metals they manage to find for in situ construction, and establish the lunar body as an easier launchpad than Cape Canaveral for flights to asteroids and other moons and planets (hence the "gas station" line). So the question from the perspective of this chapter is, *Is this okay?* Is it ethically permissible, or even existentially important, to turn the Moon into a cosmic gas station? Or does the Moon have some sort of right to remain the way it is?

Under the UN's Moon Treaty, nations are obligated to preserve the "existing balance of [the Moon's] environment."[52] But of course, none of the spacefaring nations has signed this treaty—apart from Australia, whose newly minted space agency is threatening to withdraw its commitment to the Moon now that it has the capacity to get there.[53] This means that none of the nations planning to establish outposts on the Moon has stated any commitment to stewarding its environment. To the contrary, the Planetary Protection protocols that guard celestial bodies from earthly contamination (and vice versa) seem to include the Moon under "Category I," designating it "a target body which is not of direct interest for understanding the process of chemical evolution or the origin of life." When it comes to this particular body, then, the protocols state that "no protection . . . is warranted."[54]

I've been worried about the Moon for a while now—ever since I heard that rumor in 2012 that Pepsi was planning on screening a laser advertisement on its surface. As it turns out, the rumor wasn't true, but it seems to me that the actual plan is worse: dozens of nations and corporations clamoring to surround the Moon with satellites, settle it with bases, and mine it for everything it's worth.

For a moment in October of 2020, I felt a glimmer of hope reading the Artemis Accords, a bilateral agreement initiated by

NASA and signed by its aspirational partners in the "Moon to Mars" endeavor, none of whom bothered to run the agreement by the UN's COPUOS beforehand. I remember scrolling through the section headings and finding one called "Preserving Outer Space Heritage."[55] *Finally*, I thought. *Someone at NASA is worried about preserving our poor, constant Moon.* But when I flipped to the section in question, I found that "outer space heritage" didn't mean what I thought it would mean. "The Signatories intend to preserve outer space heritage," NASA writes, "which they consider to comprise historically significant human or robotic landing sites, artifacts, spacecraft, and other evidence of activity on celestial bodies." So that's what they're interested in protecting. Not the Sea of Tranquility, not the von Kármán crater, not Mont Blanc or Mons Huygens or Promontorium Heraclides. Just the stuff that humans and robots have left behind over the last sixty years: some footprints, some landers, cameras, blast marks, that plaque, and of course, the flags. This is the same kind of anti-mineralist thinking that recognizes the sacredness of St. John the Divine but not of Uluru: if industrial humans haven't created it, it's not worth preserving. As far as the Artemis Accords are concerned, "outer space heritage" just means Earth artifacts (much of it garbage) out in space. The Moon has no heritage of its own.

The view is perhaps nowhere more frankly encapsulated than in the words of entrepreneur and space enthusiast Marshall T. Savage, author of *The Millennial Project: Colonizing the Galaxy in Eight Easy Steps*. "We can't really mess up the Moon," he assures us,

> either by mining it or building nuclear power plants. We can ruthlessly strip-mine the surface of the Moon for centuries and it will be hard to tell we've ever been there. The same is true of atomic power. We could wage unlimited nuclear warfare on the surface of the Moon, and be hard pressed after the dust had settled to tell anything had happened.[56]

But who would want to? This is the first question I'd like to ask Savage. Why would anyone want to strip-mine *anything* "ruthlessly"? And who wants to "wage unlimited nuclear warfare"? For what?

"For profit," I imagine he'd respond. "And for Progress, and the future, and the salvation of the species. Anything is worth saving humankind."

Maybe, but is Savage's promise even true? Even if we wanted to, could we really act with impunity on the surface of our only natural satellite? The answer: probably not. As space writer and anthropologist Ceridwen Dovey insists against Savage, "we *can* mess up the Moon" by polluting its fragile exosphere with rocket exhaust, and by irreversibly damaging a surface that's remained more or less unchanged for billions of years. One could list the damages that such impending lunar assaults would do *to us*, like complicating our frictionless takeoffs and landings, barring our way to our knowledge of the early solar system, etc. But again, the more difficult question is whether it's okay to do it to *the Moon*.

According to a group of Australian scholars and activists, it is absolutely not okay to do it to the Moon. They appeal to "Rights of Nature Laws," which have granted legal standing over the last few decades to natural formations like the Ganges River in India, the Atrato River in Colombia, and Te Urewera rainforest in New Zealand. In light of the escalating national and corporate effort to inhabit and exploit it, these lawyers, anthropologists, architects, and ethicists have issued a "Declaration of the Rights of the Moon."[57]

Far from being a mere object for exploitation, the declaration argues, the Moon maintains its own environment and landscapes, contains a history we are just beginning to understand, "holds deep cultural and spiritual meaning for human beings," and "is critically important to the healthy functioning of Earth." With these attributes in mind, the authors declare that "the Moon . . . is **a sovereign natural entity in its own right**."[58] As such, it "possesses fundamental rights," which chiefly include the rights not to be decimated by human pollution, extraction, or war.

Traditionally, Western law has only granted rights to beings it considers to be "persons." It is well known that for its first eighty years, the US Constitution counted enslaved people as three-fifths of a person. It sounds absurd even to have to say it, but the unfinished struggle to secure rights for African Americans requires above all a basic recognition of their full personhood. (The tragic self-evidence of this position still speaks through today's declaration that Black Lives Matter.) Activists have taken up similar strategies as they seek personhood status for sacred rivers and mountains, nonhuman animals, and, of course, corporations.

Unfortunately, only this last category has attained the unequivocal status of legal personhood under US law, relegating all prospective river persons, mountain persons, and pig persons to "resources" for private gain. "Indeed," writes Thomas Berry, a "geologian" and Roman Catholic priest, "the basic purpose of government and of the entire legal system in America has been to assist and even to subsidise the industrial corporations in their exploitation of nature."[59] The problem is, this priority of humans and their corporations over the natural world is self-defeating. As Berry writes, "we cannot have healthy humans on an unhealthy planet." The only way to secure the rights of humans is therefore to protect the rights of "those life forms on which humans most depend," including animals, vegetables, and minerals.

To someone like Zubrin, this sort of thinking is insane. The Moon, he says, has no "'right' to remain unchanged." In fact, it has no rights at all, because "clearly the Moon is a dead rock. It cannot do anything, or desire to do anything."[60]

To the contrary, "space archaeologist" Alice Gorman argues that the personhood of the Moon is far more evident than the personhood of, say, ExxonMobil. According to Gorman, the two major features of personhood are *memory*, or the "knowledge of past events," and *agency*, or the capacity to act.[61] The Moon, she says, stores memory not only in Armstrong's footprints and Aldrin's lanyards but also in its own water ice, craters, and lava fields. These formations, which Rolston would call projects, contain records of the early history of our solar system—along with hints about the evolution of others. They can therefore be understood as recollective functions. In short, the Moon has memory.

In addition to memory, Gorman argues that Moon clearly has agency. It doesn't just sit there passively, waiting to be landed on and used. Rather, the Moon creates and regulates earthly tides and stabilizes our planet's rotation on its axis. Even on its own terrain, Gorman argues, the Moon is "a very active landscape" that, among other things, "places constraints upon the desires of humans to plunder its resources."[62] My favorite example of lunar agency is the trouble the Moon gave Armstrong and Aldrin when they went to install that first flag. The engineering team had remembered there would be no wind to keep the flag aloft, but they didn't seem to anticipate the lack of welcoming soil. As space writer Christopher Potter explains, "Armstrong broke into a sweat as he struggled to push the flagpole into the ground," realizing there was "nothing between dust and solid rock."[63] If you watch the footage of the flag pole's erection, you can see that the two men spent three and a half minutes trying to get it in place—rearranging its position, manipulating its fabric, and even trying to kick some dust into a little mound to hold it aloft. And the moment the lunar module blasted off the surface, the flag fell over, giving the Moon the last word.

In addition to being recalcitrant to penetration, Gorman tells us that lunar dust is "sticky, abrasive, and full of sharp fragments of obsidian. It reacts to human disturbance by mobilizing microscopic dust that irritates lungs, breaks down seals, and prevents equipment from working properly."[64] Against Zubrin, then, Gorman shows that the Moon *does* do things. In fact, the Moon might even *desire* things. Considering the respiratory trouble it's given our astronauts and the functional trouble it's given their machines, the Moon might well be expressing a geologic desire that human beings remain on their home planet.

And we might apply similar thinking to the effort to colonize Mars. As Sparrow suggests, "If we have to wear space suits to visit and completely remodel it in order to stay," then maybe Mars is "not our place."[65]

On the one hand, the message is so clear. Obviously, Mars isn't our place; look at the lengths we'd have to go to just to have a prayer of survival there. On the other hand, it's hard to imagine the politicians, astropreneurs, and colonial nostalgics actually pausing to listen to what these celestial bodies might be saying. It would be so *inconvenient* to admit that outer space—which is clearly trying to kill us—doesn't want us there. As Earth Alliance lawyer Michelle Maloney recently said about the scramble for the Moon, "What if the answer was, 'leave the Moon alone?' No one wants to hear this."[66]

Kinship

I said a few moments ago that it didn't matter whether or not you agreed that rocks have rights, and I'll reaffirm that statement here. (Similarly, it doesn't matter whether you believe the Moon might have preferences.) For some people, such a position risks sentimentalizing itself into incoherence. For others, it risks undermining the more pressing demands of oppressed human beings and non-human animals. And for others, including Maloney, the effort to grant personhood to the Moon still affirms the (white, European) human being as the standard of a rights-bearing entity and thus as the most important thing in the world. But regardless of the answer, the *question* of the rights of rocks asks us to slow down and consider the possibility that there are some things that can't be property. And it's only by admitting that the whole universe might not belong to us that we can ask, with Rolston, how we might belong to it.

Perhaps part of the problem is linguistic. For the Bawaka people of Australia's Northern Territory, the part of the world that Westerners call "outer space" and describe as an infinite emptiness is neither empty nor "outer." Rather, the place beyond our atmosphere is Sky Country, where the ancestors live and which is subject to the same protocols of care and respect as the people's terrestrial homeland. Whenever a member of the community dies, the mourners send them through ritual song on a journey from Earth to Sky along the River of Stars—the formation Westerners call the Milky Way. And from their dwelling places among the stars, the ancestors continue to witness and influence the lives of their people.

In recent years, the Bawaka People have begun to raise objections to the behavior of the industrialized nations in space.[67] They are concerned about the disrespectful treatment to which the spacefaring nations and their partner corporations intend to subject the Moon, asteroids, and Mars—whose bodies deserve to be cared for rather than strip-mined. They are particularly concerned with the techno-garbage littering low Earth orbit, which not only obscures their vision of the stars but might also physically disrupt the passage of the ancestors from Earth to Sky. Far from being a frontier to conquer, the Bawakan heavens safeguard the lifeworld of Earth. So who would want to ransack the heavens? (Well, we might answer, the same people who've ransacked Earth.)

Although it's exemplified by the Bawaka People, this notion of a celestial kinship is hardly limited to them. The Navajo Nation refers to the heavens as Father Sky. The Skidi Pawnee teach that the people of Earth were conceived by the stars. The Zuni people call the Sun "Father" and Earth "Mother," and the stars are their other relatives. Christians in the Franciscan tradition call the Moon "Sister" and the Sun "Brother." And like the ancient Greeks, contemporary pagans affirm that the Moon is the realm of the goddess Artemis.

During the summer of 2020, some "baby witches" (that is, young and inexperienced practitioners of Wicca) tried to hex the Moon on TikTok. Understandably, this action invoked the rage of their pagan elders on behalf of Artemis and her twin brother, Apollo, who, incidentally, is the god of healing.[68] "How dare they mess with powers they don't understand," fumed the elders. "Especially considering the epidemiological mess we're in on Earth—you'd think they'd want to appease the god of healing, rather than enrage him." Just wait till the witches hear what NASA's planning to do to the Moon—on a mission they've named after Artemis.

* * *

In the early 1970s, an anthropologist in Alaska asked the Inuit people she was living with whether they'd heard about the Apollo missions. She told them about the first rocket to orbit the Moon and the first ship to land there. She told them about Armstrong and Aldrin and the flag and the footprints. Then, as another scholar retells it,

> The Inuits began laughing, and when the anthropologist inquired why, they replied: 'We didn't know this was the first time you white people had been to the Moon. Our shamans have been going for years. They go all the time. . . . The issue is not whether we go to visit our relatives, but how we treat them and their homeland when we go.[69]

Again, you don't have to hold any particular beliefs about the matter to spend time with the question. Suppose the Moon *were* the place "our relatives" lived—how would we treat its surface, its ice, its atmosphere? How would we treat the spaceways if we believed that our loved ones lived there after death? And how would we treat Mars if we thought it was a legal person—or a god, for that matter? The question isn't whether these things are true but how we'd behave if we thought they were. As the pagans might ask, *What would Artemis mine?*

CHAPTER SEVEN

Other Spacetimes

If an angel led the Exodus out of Egypt,
why can't an angel lead some others somewhere else.
SUN RA

Where Is Everybody?

There's a riddle in astrobiological circles called the Fermi Paradox, which goes like this: there are a billion trillion stars in the observable universe. That's 100,000,000,000,000,000,000,000 possible solar systems, each orbited by anywhere between zero and eight (or maybe more) planets. Even if only one in a thousand of these planets is inhabited—even if it's only one in a million or one in a billion—there's got to be an absolute ton of life out there, and a good deal of it should be older and more technologically advanced than earthlings. So, as the Italian physicist Enrico Fermi famously asked, *Where is everybody?*

Recent years have seen the declassification of information on "unexplained aerial phenomena," or UAPs, the Pentagon's new name for those sightings formerly ridiculed as UFOs.[1] As it turns out—at least if we can trust the testimony of thousands of pilots, military operators, astronomers, and ordinary citizens—there seems to be all sorts of stuff flying around out there. It's still unclear whether the stuff is terrestrial or extraterrestrial, organic or technological, mechanical or intelligent, or some combination of all these things. But in the meantime, the thousands of cataloged UAPs only intensify the Fermi Paradox: if the universe is filled not only with aliens but with aliens who know how to find us, then why haven't they revealed themselves?

"Maybe they don't like the way we're behaving." Afrofuturist curator Ingrid LaFleur said it offhandedly, in an online seminar to which astrobiologist David Grinspoon and I had invited her in the spring of 2021. "We've already disrespected the cosmos in our efforts to explore space. . . . All we do is leave millions of pieces of junk behind. I can see why they wouldn't want to have anything to do with this planet."

"Ha!" Grinspoon exclaimed. "That might be the best resolution to the Fermi Paradox I've ever heard! 'Why aren't they coming? . . . Because they've seen what we're like and they don't *want* to!'"

Again, you don't have to hold any particular beliefs about this issue to spend some time with the questions it poses. There may be life beyond Earth or there may not. Such life might conduct periodic flybys in disk-shaped ships whose motion contradicts the laws of physics or it might not. Regardless of whether such beings exist, however, it is instructive to ask what they'd see if they did. What kind of galactic and universal citizens would these hypothetical aliens think earthlings to be—what with our warming planet, warring nations, swirling garbage, and aspirational cosmic gold digging?

I'm not sure how to answer my own question. As far as I know, the aliens haven't contacted me. But I imagine that intelligent beings from other worlds would respond to our incursion into space much like the First Nations responded to Europe's crossing of the seas: with some mixture of curiosity, wonder, incomprehension, disgust, and fear. It's a good bet that the response would be the same because the strategy is the same: send a few advance missions, establish outposts, reconfigure the land, extract resources, sell resources, build colonies, and back the whole thing up with cannons and soldiers.

As we have seen, the corporate space enthusiasts insist that the game is different this time because the lands they're aiming for aren't inhabited. The Europeans *mistook* Africa, the Americas, and Australia for empty land, and they committed unthinkable violence against the land's Indigenous caretakers. Even the most ardent of terraformers will at the very least say that the enslavement and destruction of First Nations was wrong. But when it comes to space, the terraformers insist, there's actually nothing there. In space, we can finally feel good about frontierism because we've finally got an empty frontier.

Now just as I did regarding the personhood of rocks, I'm going to ask you temporarily to suspend judgment when it comes to extraterrestrials. For the moment, let's pretend it doesn't matter whether or not aliens exist. Let's ask, instead, whether the colonial approach to space is beneficial to the beings who most certainly do exist, here on Earth.

According to the major space actors, the answer is a resounding *yes*. The American-led, corporate-backed conquest of space will benefit all humanity by ensuring that democratic values govern space (JFK), by protecting the military and economic interests of the US and its allies (Space Force), by preserving Western civilization (Zubrin), by extending the light of consciousness (Musk), by solving the energy crisis (Bezos), and by letting ordinary people gaze back at Earth from space (Branson). Even the mining companies have taken up this humanitarian argument, insisting like the CEO of Moon Express that they are turning the lunar body into a gas station "for the benefit of humanity."[2]

But just as Neil Armstrong's Moonwalk has done very little for poor, Black, Indigenous, and immigrant people in his own country—not to mention all "mankind"—the pursuit of profit in space will leave noninvestors even further behind than they already are. As the *New Republic*'s Clive Thompson predicts, "the big winners in space will likely be . . . the big winners on Earth."[3] After all, a corporation's chief obligations are to its wealthy shareholders—not to its workers or even its clients, let alone all of "humanity." And yet the corporations keep feeding us these mystifying promises to "benefit humanity," assuring us, in the words of science writer Martin Robbins, that "when we go into space, we will all magically become nice."[4]

Under these conditions, though, it's just so unlikely. Do we really expect that the billionaires who can't find any cause worth supporting on Earth will finally redistribute their wealth once they get deeper into the final frontier? Do we really expect that the notoriously inhumane industries of mining, manufacturing, and global retail will suddenly establish decent working conditions on literally uninhabitable planets? And what about all the ecological damage they're doing in the meantime on Earth?

What happens to our environment when a single rocket scorches the land it leaves behind, drops its boosters in low Earth orbit or dumps them in the sea, and deposits millions of pounds of rocket fuel into the atmosphere? What happens to the Boca Chica residents SpaceX displaces from their suddenly toxic homes, including the humans who now can't afford to live anywhere else and the wildlife that increasingly has nowhere to go?[5] In what sense is this adventure benefiting all humanity?

A large part of the problem lies in the word *humanity* itself. I'm hardly the first to have had this thought; Caribbean, African, and African American philosophers like Frantz Fanon, Sylvia Wynter, Achille Mbembe, Charles Mills, and Saidiya Hartman have long exposed the term as a dangerous abstraction. In the guise of inclusiveness, the language of "humanity" has traditionally subjugated nonwhite, non-European subjects to European rule. In particular, as Wynter shows, Europe swallowed the Americas whole with its allegedly universal humanitarianism.[6]

Armed with the belief that all humans were made by the same God and saved by his same son, the Christian nations justified every imaginable violence as necessary means toward the holy end of saving souls. Similarly, the astropreneurs shake off all ethical critique by saying that environmental degradation, obscene hoarding of wealth, and neglect of actual human suffering are all necessary means toward the holy end of saving the species. But if these galactic messiahs kill off most of us in the process—along with the only biosphere that allows us to *be*—then what, exactly, are they saving?

The operative fallacy here is known as *longtermism*. Popular among Silicon Valley types, the idea is that the galactic immortality of the species is more important than the current well-being of any existing community. Hunger, poverty, racism, warfare, hurricanes, floods, pandemics, and extinctions might *seem* like enormous concerns, but "from the perspective of humanity as a whole," they are just the ups and downs of the evolution of the species. As techno-philosopher Nick Bostrom assures us, "even the worst of these catastrophes are mere ripples on the great sea of life."[7] So the key is to rise above the everyday struggles of particular humans and focus instead on the long-term existence of the species.

I'll give away the punchline and say that longtermism is a high-tech version of what Malcolm X called "pie in the sky and heaven in the hereafter." Sick of the suffering Black Americans continued to undergo for the sake of an unjust law and order, Malcolm blamed America's racist social system on the Christian teaching that earthly suffering would be rewarded in the afterlife. To his mind, the doctrine of heaven maintained and even glorified oppression, convincing Black Americans that it was useless or even ungodly to overthrow their oppressors. From the perspective of eternal bliss, they were taught to believe, poverty, racism, and even enslavement would seem like nothing at all—*mere ripples on the great sea of life*.

Once again, then, we find the astropreneurs selling a Christian doctrine that even the Christians have abandoned. Since the end of the second World War, the idea of redemptive suffering and the single-minded focus on heaven have disappeared from all but the most conservative theologies.[8] For the most part, the Christians are out feeding the hungry, clothing the refugees, visiting the sick and imprisoned, hosting the recovery meetings, and marching in defense of Black Lives. Meanwhile, the billionaires are holding out for heaven on the asteroid belt.

There's a murderous numerology at work here that sets the eight billion people on the planet (seven hundred million of whom live in extreme poverty) against, say, the 10^{23} people who *might* exist if we manage someday to colonize the Virgo Supercluster.[9] The number 10^{23} is greater—a whole lot greater—than eight billion or seven hundred million; therefore, the long-termers insist that our energies be directed toward the hypothetical humans rather than the actual ones. In fact, they caution, actual humans might not be actual at all. They could be computer simulations.

I'm not making it up. The idea's been around in one form or another since René Descartes, but it came raging back in 1999 with *The Matrix* and in the gamer-geek philosophy of Bostrom, who argued four years after the Wachowskis introduced us to Morpheus that if it is possible (even in principle) to create a conscious simulation, we are almost certainly living in one.[10] In the meantime, the futurist economist Robin Hanson reasoned that if we *are* living in a simulation, we should behave as brashly and boldly as possible so that our simulators remain sufficiently entertained to keep us plugged in. Assuming that we are living in a simulation should have the overall effect, Hanson agreed, of making us more present oriented and more selfish than we might otherwise be. "Your motivation to save for retirement," he reasons, "or to help the poor in Ethiopia, might be muted by realizing that in your simulation, you will never retire and there is no Ethiopia."[11]

I won't drive you mad with the details of these scenarios, which are mind boggling at best and diabolical at worst. I bring them up only as a means of filling in the intellectual biography of our most energetic space utopian, whose deep dive into longtermism has allowed him to propose his *own* solution to the Fermi Paradox.

"If an advanced system existed at any place in this galaxy," asks Elon Musk, "at any point in the past 13.8 billion years, [then] why isn't it everywhere?" As Fermi asked, *Where is everybody?* Musk's answer: we're probably living in a computer simulation. Our consciousness is probably the creation of superintelligent beings running a superadvanced version of Sim City, and the rest of the universe is just a virtual backdrop to our trivial pursuits. We've therefore got to be as daring as possible, earning extra lives and power-ups to unlock new worlds before the simulators end the game.

Or it's always possible that we're not so much simulated as manufactured. "If it's not a simulation," Musk speculates, "then maybe we're in a lab and there's some advanced alien civilization that's just watching how we develop, out of curiosity, like mold in a petri dish."[12] But whether we're mold spores or Mario and Luigi, the quest is clear: to stay alive by going forth, increasing, and multiplying. And here's the thing: if the world around us is just a quantum computation or an alien petri dish, then who cares about the wetlands or the coral reefs? If they're really important, the simulators will make more. And if not, the herons, frogs, and great sea turtles will have been a necessary trade-off to get humanity up to the next level on Mars.

You may have noticed that in the midst of all this quantitative thinking, the qualitative disappears completely. Our minds become so dazzled at the thought of a scadzillion descendants on a million planets in a high-tech multiversal simulation that we forget to ask, not so much what kinds of lives the rich guys' progeny will lead but what kinds of lives the rest of us will continue to live in order to sustain this overgrown gamer fantasy.

"A Burning Need for Something Else"

In "The Ones Who Walk Away from Omelas," science fiction author Ursula Le Guin describes a gleaming utopian city. The streets are bright and colorful, the parks healthy and plentiful, the people merry, the music gay, the animals well cared for, the sun and rain in perfect balance, and the technology sufficient to everyone's comfort within the bounds of mutual care (e.g., there are "very fast little trains and double-decked trams," but no cars or helicopters).[13] The only hitch is that the joy of this city is sustained by the unrelenting suffering of a single child—naked, sporadically fed, and locked in a basement with old mops and dirt floors.

"They all know it is there," the narrator explains,

all the people of Omelas. Some of them have come to see it, others are content merely to know it is there. They all know that it has to be there . . . that their happiness, the beauty of their city, the tenderness of their friendships, the health of their children, the wisdom of their scholars, the skill of their makers, even the abundance of their harvest and the kindly weathers of their skies, depend wholly on this child's abominable misery.[14]

In a brief preface to this story, Le Guin traces its central conceit back to the American psychologist, William James, who offered it as a thought experiment. What if millions of people could flourish forever "on the one simple condition that a certain lost soul . . . should lead a life of lonely torment"? James finds the prospect unbearable. None of us, he insists, would accept such a bargain. In fact, the very knowledge of the torment of that one soul would ruin the happiness of all of our millions—wouldn't it?

Dear old William James, smiles Ursula Le Guin. "So mild, so naively gentlemanly. Look how he says 'us,'" she marvels, "assuming all his readers are as decent as himself!" But of course, we are not. What Le Guin's narration allows us to realize is not only that *some* people are happy to balance their happiness on the suffering of others—like the Omelans and their sacrificial child—but that most of "us" are as well. Most of the wealthy residents of the overdeveloped nations are delighted to find bargain clothes made by children in sweatshops. Most of us are happy to travel to resort towns whose residents are denied access to their own beaches. Most of us love fresh fruit even though it's been grown and picked under unbearable conditions, flown on jets that pollute the airways, driven on trucks that kill whatever tries to cross the road, and carefully stacked by people who can't afford to buy it. This is the way capitalism works: the happiness of some is sustained by the suffering of everyone else. The only difference between us and the Omelans is that we don't really want to know any of this—and of course, that the suffering children we rely on are horrifically more numerous than one.

Le Guin's story ends with a tribute to "the ones who walk away," those few Omelans who realize that their happiness isn't worth the suffering of even one child and who walk alone into the darkness. "The place they go towards is a place even less imaginable to most of us than the city of happiness," the narrator confesses; "I cannot describe it at all. It is possible that it does not exist. But they seem to know where they are going, the ones who walk away from Omelas."[15] And the question is whether you, the reader, are able to see it too. Can you imagine a society that doesn't depend on the misery of others? Can you see it clearly enough to believe it, and can you believe it enough to start walking there? Can I?

"But where would we go," you might ask. Given the global, and now outerspatial reach of corporate capital, there is no "away" left—at least not one to which any of us has physical access. Where could we possibly go to live otherwise? Even if we could find such an elsewhere, what about that poor, naked, suffering child? How would our walking away from Omelas do anything to change Omelas? Couldn't we be accused like the rich guys of blasting away from our world rather than caring enough to fix it?

These are perhaps some of the questions Afrofuturist sci-fi author N. K. Jemisin had in mind when she wrote her narrative response to Le Guin titled, "The Ones Who Stay and Fight." In the resplendent city of Um-Helat, the people are well fed, healthy, even joyous. They hail from diverse backgrounds, speak multiple languages, and enact laws that ensure the safety and comfort of all citizens. "And so this is Um-Helat," the narrator explains: "a city whose inhabitants, simply, care for one another."[16]

Throughout the story, the narrator pauses to address our likely skepticism. How could a society possibly be this functional, this communally minded, this *joyous*? Surely there's a hitch? Of course there's a hitch. And much like Omelas, the hitch of Um-Helat involves a child. A single child whose father has discovered transmissions from the "benighted hellscape" of the readers' own world and who has begun to learn and spread the idea that *"some people are less important than others."*[17] A child whose father's new convic-

tion threatens the society so fundamentally that a panel of reluctant social workers finds him, kills him, and then turns to face his
enraged, grieving, ideologically poisoned daughter. And just as the
reader braces herself for more violence, one of the social workers
"crouches and takes the child's hand."[18]

"What?" the narrator exclaims. "What surprises you? Did you
think this would end with the cold-eyed slaughter of a child? There
are other options—and this is Um-Helat, friend, where even a piti-
ful, diseased child matters."[19]

There are other options. Not somewhere out there, but in the mid-
dle of the mess we're in. Finding them and living them out would
be the work of the ones who stay and fight. In other words, the ones
who stay are actually the makers of other worlds. Meanwhile, the
escape artists refuse to see that the world might be otherwise. The
stayers make other worlds while the leavers preserve the old one,
refusing to see any other options.

I point out this ironic alignment not in order to raise objections to
Le Guin but to point out the deeply conservative nature of the astrotopians. The reason they want to leave Earth is that they want to
keep the world the way it is. So the rocket men seek more land and
resources to plunder in space. Meanwhile, the hungry, colonized,
Black, Indigenous, working-class, Earth-loving, and peace-seeking
among us are filled with what Sun Ra called "a burning need for
something else." Not the same system in some cosmic future but a
radically different system right now and right here. As the Ameri-
canist Jayna Brown writes in *Black Utopias*, "we must jump into the
break, the cut, into an entirely different paradigm."[20]

Art, Perhaps

An entirely different paradigm! But where are we going to find one?
Paradigms are the filters through which we understand the world.
So how can we imagine beyond the stuff that allows us to imagine
in the first place? It's like trying to dream up a brand-new color or a

mythical creature that isn't just a mash-up of other creatures. How can we think the unthinkable?

Our old friend Nietzsche found himself in a similar ditch in his critique of Christianity. As you may remember from our exploration of colonialism's biblical legacy, Nietzsche calls modern science the inheritance of ancient Christianity, which sets us against our own pleasures, our well-being, and Earth itself. Everywhere he looked, Nietzsche saw this "nihilism" at work: in religion, of course, but also in the sciences (and he didn't even live to see the atom bomb), philosophy, language, politics, and economics. All of the dominant, Western ways of knowing keep us docile in the face of our steady self-destruction.

So what can we do instead? Nietzsche was at a bit of a loss. After all, the same systems that had made nihilists of the rest of the world had made one of him too. And yet, he was certain that there were other ways of being—ways that could say "yes" to life and energy and flourishing—he just couldn't see them anywhere. So the trick would be to make them up.

To make them up. Not to discover them, like an explorer or scientist or pilgrim would try to do, but to *create* them. Brightened by this hope in creativity, Nietzsche momentarily suggests that art might finally get us out of nihilism—before grumbling about how easily corrupted artists tend to be.[21] And it's true. Art, like everything else, is more liable to reflect the status quo than it is to change it. But some of it might actually slip through, allowing us from the depths of destruction to imagine genuinely other worlds.

This was the message Le Guin herself delivered when she received the 2014 National Book Foundation Medal for Distinguished Contribution to American Letters. It was an unprecedented award for an author of science fiction, a genre that "realist" critics tend to dismiss as juvenile and escapist. "Hard times are coming," Le Guin said two years before the election of Donald Trump, "when we'll be wanting the voices of writers who can see alternatives to how we live now." According to Le Guin, it is authors of speculative fiction, science fiction, and fantasy "who can see through our fear-stricken society and its obsessive technologies to other ways of being, and even imagine real grounds for hope."[22] Other ways of being—as in Jemisin's Um-Helat, whose inhabitants can all have homes if they want them.

Granted, most fantasy and science fiction doesn't work to imagine genuinely different worlds. Most of it replays the imperial story that Diné (Navajo) critic Lou Cornum calls "white male explorer . . . meets exotic, primitive, mysterious, threatening alien."[23] But some of it—especially in the hands of racially and sexually marginalized artists—manages to take this story and turn it on its head, or on its side, or to rearrange it completely so that the alien territory becomes a place of genuine liberation rather than liberation at the expense of Indigenous people, exploited laborers, and the land itself.

The genre of Indigenous Futurism, for example, recovers the myths, rituals, and storytelling of First Nations to build different kinds of worlds than the one that has taken over the globe. As Anishinaabe scholar Grace Dillon explains, "the Native Apocalypse, if contemplated seriously, has already taken place."[24] The voyages of Columbus unleashed the literal end of the world for many First Nations, who have not only learned to survive the apocalypse but to create new ways of being from the mess it's left behind. For example, Cornum calls our attention to the novel *Ceremony* by Leslie Marmon Silko (Laguna Pueblo), whose most adept medicine man bundles together traditional herbs with trash from the "dumping ground" he's living in to heal his community's psychic trauma.[25] Other options—not in some untainted elsewhere but right in the midst of the mess.

Or there's *The 6th World*, a short film about a joint Navajo-NASA mission to Mars. When the space agency's sanctioned GMO corn crops all die, Navajo astronaut Tazbah Redhouse germinates the ancestral corn pollen she's smuggled into the spaceship and saves both herself and her non-Native colleague. "One of the most powerful narratives offered by Indigenous Futurism," Cornum explains, "is that we Indigenous peoples are carriers of advanced technical knowledge that can be applied in ways much more profound and generative than the extractive, destructive, life-denying processes of capitalism and Western progress."[26]

One of the subtler violences of colonialism is its obliteration of all stories and ways of knowing other than its own. By cutting enslaved and Indigenous people off from their land, their languages, and their rituals, ancestors, and gods, Western imperialism has presented itself as the only living history and therefore as the only possible future. We come to believe there is simply no alternative to Western values, technology, politics, and economics. (Even my students tell me it's more likely they'll live in a rotating corporate space pod than a just society on Earth.) But, to put it bluntly, there are all *sorts* of alternatives to Western technology, politics, and economics. The task, as Indigenous Futurism shows, is to recover the deliberately erased histories of colonized and oppressed peoples, sift through them to find the most life-giving stories, and bundle them together with the mess of the modern world to imagine new ways of being.

Indigenous Futurism inhabits multiple times at the same time to open what I'm calling "other spacetimes." As such, this transformative genre is indebted to its older sibling Afrofuturism, which British Ghanaian theorist Kodwo Eshun calls "a program for recovering the histories of counter-futures."[27] The term *Afrofuturism* tends to be attributed to the Anglo-American critic Mark Dery, who proposed it in conversation with Black sci-fi authors Samuel Delany, Greg Tate, and Tricia Rose.[28] And although some artists resist calling themselves Afrofuturists because of the name's white, American, and largely male origins, others have expanded, revised, and refined it in relation to their own transformative visions.[29]

Perhaps the best-known Afrofuturist is the experimental jazz musician Sun Ra. Having abandoned the "slave name" he'd been assigned at birth, Ra remade himself in relation to Egyptian divinity, Ethiopian royalty, and the vibrations of other planets. Disgusted by the system that had enslaved Black Americans, Ra connected a noble African heritage to the other worlds of outer space, which he often presented as the site of his own past. Time folds in on itself here, as Ra rejects the Western march of empire, touted as "progress," from ancient Rome through modern Los Angeles. Space does not lie in a simple future any more than "Africa" lies in the past. Rather, all of it contributes simultaneously to Ra's imagination of totally different spacetimes right where we are.

"These space men contacted me," Ra explains; "they wanted me to go to outer space with them. They were looking for somebody who had that type of mind."[30] Through a process he calls "trans-molecularization," Ra says that the aliens changed his whole body "into something else," brought him to Saturn, and then deposited him back in Chicago to instruct earthlings in the ways of higher consciousness. At other times, Ra would say he was actually *from* Saturn, his home planet, where he had learned the music and philosophy of "superior beings."[31]

Ra's task? To take a traditionally dehumanized people and lift them—not into "humanity," but beyond it. Like many of his Afrofuturist descendants, Ra has absolutely no interest in the category of humanity, which as we have seen has historically been denied to everyone but free, European-descended bipeds. "Man is a horrible word to galactic beings," Ra said; "my job is to change five billion people to something else."[32] This "something else" would be utterly unlike the selfish, individual, Western "human" who had strangled his own planet. Perhaps, Ra mused, a "living cosmic multi-self."[33]

But even this language might make Ra's postself sound too bounded, too possessive. What Ra was really trying to get folks to do was to lose themselves entirely. "Ain't nobody there," Ra wrote of the cosmic multiself; "it's just a hyper dimension of a / concurrent coplanar force that / tends to make you / part of the universe."[34] And how do ordinary folks go about reaching this hyper dimension? How can we lose ourselves to become part of the universe? Through music, of course.

Take a minute, if you would, to listen to "We Travel the Spaceways"—especially as it's recorded on Sun Ra's *Greatest*

Hits: Easy Listening for Intergalactic Travel. You can find it easily online.[35] You'll hear a low piano riff that sounds like an old steam train is chugging along the tracks, punctuated by arresting chords and a cymbal. Numerous voices sing a deceptively simple melody crowned by a train-whistle harmony, announcing "we travel—the spaceways—from planet—to planet." The piano draws you into its uncompromising rhythm and playful, midrange splashes; it's impossible not to find yourself rocking along on this cosmic railroad ride. Some trumpets slide in, announcing arrivals and departures.

Gradually, Ra starts adding in crowded chords—nothing atonal, just clashing enough to make you pause and wonder if you heard them right. Suddenly, a tenor saxophone flies in in a totally different mood and range, just missing the high notes on purpose, trilling here and there for the fun of it, modulating the melody, and allowing Ra's riffs to grow less and less predictable as the saxophone gives ways to screeching before the whole thing settles back into the piano's repetitive bass line again, the muted trumpets return, and those reassuring voices remind us that we're traveling the spaceways from planet to planet.

"Superior beings definitely speak in other harmonic ways than the earth way because they're talking something different," Ra explained. "And you have to have chord against chord, melody against melody, rhythm against rhythm: if you've got that, you're expressing something else."[36] The music, in other words, *is* the message. It's also the medium: the means by which listeners are transported from this unbearable world into another dimension, another spacetime, "another kind of living life."

And so Ra called his band the Omniverse Arkestra. The music itself was the ship—the *ark*—that could transport the faithful remnant out of this doomed world and into all possible worlds of the "omniverse." In addition to Noah, Ra often playfully compared himself to Moses. "If an angel led the Exodus out of Egypt," he reasoned, "why can't an angel lead some others somewhere else."[37]

And it's here that we can appreciate the intensity of Ra's counternarrative—of his telling the same story differently. Unlike the American founding fathers, who also leaned heavily on the Exodus narrative, Ra isn't leading his people out of earthly oppression so they can become cosmic oppressors. He never speaks of Venus as Canaan, Saturn as Zion, or the galaxy as a frontier. Nor, at the risk of pointing out the obvious, does he actually build a spaceship. Rather, the Arkestra offers a different kind of interworldly transport, one not to conquer the spaceways but to find "another kind of living life" among them. Because the many worlds of the omniverse aren't there for us to take. They're there to take *us*. To teach us harmony. And that's the new world: another spacetime, right where we are.

Taking Root among the Stars

As we might recall, American politicians have long likened the journey through space to the journey across the oceans. Just as "our" pioneering ancestors braved the Atlantic passage to settle in a new world, so will "our" pioneering children brave the threatening expanse of outer space to plant new worlds on the Moon, Mars, and beyond. We might also recall the military dimension to this heroic retelling; as both Lyndon Johnson and JFK warned the electorate, the nation that controls space will control our future, much as the nations that controlled the seas controlled the past. Finally, we might recall that this comparison between our astronauts and our ancestors steadily leaves out those ancestors whose lands were stolen in the Americas or who were themselves stolen from Africa.

Afrofuturists take this analogy between the seas and space and reveal its sinister core. Rather than assuming the perspective of the bold explorer at the bow of a ship, they take us to the hellish hold beneath it, crowded with bought and sold bodies transported against their will. As Eshun explains, Afrofuturists see the transatlantic slave trade as nothing short of an alien abduction.[38] Otherworldly bipeds descended out of nowhere wielding weapons no one had seen before; poked, prodded, and humiliated their West African hosts; and then made off with thousands of their people to study, enslave, and assimilate. Much like Indigenous Americans, African-descended peoples already know how to survive the apocalypse and live on other worlds.

Afrofuturist author Octavia Butler sets her Earthseed series at the end of the world the colonizers have created. Written in the early 1990s, the first book is set in a midapocalyptic 2024. The United States is unraveling thanks to the escalation of climate change, infectious disease, and income inequality. Those who still have money are living in walled enclaves surrounded by unemployed thieves addicted to a designer drug that parlays the vision of fire into erotic pleasure.

When arsonists destroy Lauren Oya Olamina's family and community, she gathers her few remaining neighbors and leads them north through the industrial wilderness of postapocalyptic California. As they begin a new life together in a small, intergenerational, multiracial commune named Acorn, Olamina gradually introduces her people to the philosophy-religion she calls Earthseed. "The only lasting truth / Is Change," she teaches them; "God / Is Change."[39] According to this postapocalyptic theology, the fact of change is inescapable, but the character of that change is not. It is therefore the duty of Earthseed to shape the changes that will allow them to live and flourish. In so doing, they will be shaping God.

As *Parable of the Sower* gives way to its sequel, *Parable of the Talents*, a right-wing Christian nationalist movement takes hold of a faltering US. The "Christian America" party besieges all "heathen" outposts, including Olamina's community in the mountains of Humboldt County. For the next year, Acorn's adults live as slaves at this new "Camp Christian" while their children—including Olamina's infant daughter, Larkin—are sent to reeducation camps and adopted by Christian Americans.[40]

After an excruciating enslavement and an almost superhuman revolt, Olamina manages to drive out the Christians and reach an agreement of nonintervention and nonretaliation. She searches for Larkin but is told all records of her daughter's relocation have been destroyed. Meanwhile, Olamina sends missionaries to convert those whom the apocalypse has left behind and prepare them for Earthseed's final "Destiny": "to take root among the stars."[41]

It's a vision Olamina has had since she was a teenager: the people of Earthseed will either become "smooth-skinned dinosaurs" on a burning planet or they will fulfill their Destiny (always capitalized) in the heavens. Heaven in the actual heavens. Not on the Moon, she explains to one of her early followers, or even on Mars, but in "other star systems. Living worlds."[42] So the whole point of amassing followers and funding on Earth has been to prepare her people "to leave the protection of the mother" and set off for other worlds.[43]

We learn the details of Earthseed's astrotopianism through the voice of Asha Vere, Olamina's renamed daughter, fully grown and filled with rage. Asha Vere learns her own story slowly, accidentally, and when she finally locates her mother, she can't escape the sense that Olamina could very well have found her if she'd tried a bit harder. But as Asha Vere fumes, "Earthseed was her first 'child,' and in some ways her only 'child.'"[44]

In the prefaces that the lost daughter appends to her mother's journals, it becomes clear that her distrust of Earthseed stems from more than simple jealousy. Even before she knew she was the prophet's daughter, Asha Vere took offense at Olamina's cosmic escapism—its effort to solve earthly problems by flying off to the Crab Nebula Alpha Centauri. "To tell the truth," Asha Vere writes, "the more I read about Earthseed, the more I despised it. So much needed to be done here on earth—so many diseases, so much hunger, so much poverty, such suffering, and here was a rich organization spending vast sums of money, time, and effort on nonsense. Just nonsense!"[45] It's a criticism we might recall from Ralh Abernathy's and Gil Scott-Heron's protests against Apollo 11, Wendell Berry's invective against Gerard O'Neill, or any number of Twitterslams against Bezos and Musk: can't they spend all that money and brilliance on Earth, instead? It's a criticism we find in Jemisin's gentle revision of Le Guin and in Sun Ra's this- otherworldliness, transporting the listener through harmonics rather than hydrogen.

By undercutting Olamina's astrotopianism with Asha Vere's

reactionary disgust, Butler presents her reader with a considerable dilemma. Is Olamina a savior or an entrepreneur? Is she a prophet or an escape artist?

Right before her sudden death at eighty-one years old, Olamina sees the first space shuttles taking her followers off the planet, heading into the lands she has promised them. They will join a larger starship "assembled partly on the Moon and partly in orbit"[46] and then head to other solar systems, where some of them will take root and some of them will not (much like the sower's seeds in the Gospel of Luke). The name of the starship? The *Christopher Columbus*.

"I object to the name," Olamina writes. "This ship is not about a shortcut to riches and empire. It's not about snatching up slaves and gold and presenting them to some European monarch. But one can't win every battle.... The name is nothing."[47]

But is a name ever nothing? From her daughter's perspective, Olamina's messianic dreams are much closer to these old imperial visions than she'd like to admit. From her daughter's perspective, Olamina's concession to *Columbus* is the culmination of her lifelong concession to ego, escapism, and even conquest—as encapsulated in her fervent belief in (Manifest?) Destiny. Will Earthseed really build worlds of equality and peace? Or has Olamina-Moses, dying on the Earth-side of the Jordan, sent off her people to conquer an infinite Canaan?

It may be that Butler herself isn't sure. She spent years trying to write the final volume of the Earthseed series, which she intended to call *Parable of the Trickster*, and which she set on an exoplanet called Rainbow. But as Jayna Brown has found in her archival papers, Butler couldn't figure out what this new life beyond Earth would look like, growing so frustrated that she called the failing manuscript "a piece of garbage."[48]

Brown attributes this impasse to a contradiction between Olamina's belief in Destiny and her trust in Change. If Change is, in fact, the supreme force in the universe, then there is no such thing as Destiny—no predetermined fate for chosen humans to pursue. The novels also seem torn between the promise of other worlds and the importance of this one; as Asha Vere writes, "If my mother had created only Acorn, the refuge for the homeless and orphaned. . . . If she had created Acorn, but not Earthseed, then I think she would have been a wholly admirable person."[49] Olamina knew how to heal the land, build community, strengthen spirits, and make something beautiful out of an unbearable situation. So why did she send all that beauty into space?

Will Earthseed plant just and loving communities throughout the galaxy? Or is its imagined destiny too bound up with frontierism, messianism, missionizing, and conquest to live up to the promise of Change? Will the pioneers shape a God who shapes them into something more peaceful than "humanity"? Or will they shape God in all the old, imperial, dominion-giving ways? Ultimately, Butler's reader isn't quite sure what to think. In the end, *Talents* leaves us with Olamina looking up to the stars she'll never reach, saying, "I know what I've done." And we're free to read that declaration as triumphant, ominous, or both. What's important is that we ask where and how the real change might happen—whether this new God's new world is lying in wait somewhere out there, or whether it might already be happening in the ruined Earth.

Imagine Dragons

This emerging thread of earthly otherworldliness—of other spacetimes opening in the midst of wherever it is we are—finds tragic elaboration in Jemisin's "Cloud Dragon Skies." As do the *Parable* novels, Jemisin's story opens on an increasingly uninhabitable Earth: "all the world blew poisons into the sky. Forests died. The world grew warmer."[50] An omnifirm called Humanicorps consolidates profits on a dying planet while collapsing the political order. And finally, the disaster gives way to a "great exodus" into outer space.[51] But unlike Butler's Earthseed or Ray Bradbury's *Martian Chronicles* (which imagines a great "river" of African Americans getting the hell off this planet), Jemisin's world is one in which the oppressors leave and the oppressed stay.

It's not quite that stark—nothing in Jemisin is—but despite most characters' moral complexity, the exodus presents each of them with two options. They can move to "the Ring," a corporate-run asteroid colony "where there could be cities and cars and all the conveniences of life as it once was," or they can stay on Earth and have "nothing."[52]

The whole story is basically a deep dive into Jeff Bezos's rejected counterargument about "rationing" and "stasis." As you may remember, Bezos reasons that we could refrain from colonizing space and mining its resources, but who would want to live a "dull" life of low-energy use and subsistence production? Jemisin's character Nahautu raises her hand. "Most chose the Ring," she explains; but she, her father, and many others chose to stay on their ruined Earth. And almost immediately, as the last comfortable capitalists blasted off from the spaceports, all the ways of living they'd suppressed came back. The Buddhist, Jewish, and Indigenous sages "came forth and taught the people anew all the ways they had once scorned. And all the clans everywhere, no matter their chosen ways, swore the same oath: to live simply."[53]

So that's the only rule governing the postapocalyptic Earth: to live in accordance with nature rather than the other way around. "We no longer change the world to suit ourselves," Nahautu's father explains; "when the world changes, we change with it."[54] Already, then, we can see the reversal of Olamina's Earthseed theology. Rather than trying to control the shape of Change, the earthly remnant is letting Change reshape them.

The most significant of these changes—more so even than the disappearance of allegedly advanced technology, along with the majority of humanity—is the color of the sky. Shortly after the exodus, toxins that had lain "dormant" in the soil and the atmosphere "awakened, combined in some strange new way, and changed the sky."[55] Rather than blue with puffy, floating clouds, the atmosphere is now pink with spiral, diving mists. To Nahautu, the spirals look like dragons: protective, threatening—maybe both.

One day, a small delegation of colonists from the Humanicorp habitat return from their asteroid ring to Earth, convinced they've discovered a way to fix the sky. Nahautu's father protests that all earthlings have pledged not to tamper with their world anymore but rather to accept what it gives them and adapt accordingly. The sky people ignore him, releasing a single missile into the air to "neutralize the chemicals."[56] A sudden blue spreads through the sky for a few moments, until thick thunderclouds gather, lightning blazes in every direction, and the infuriated cloud dragons attack the bedraggled Earth.

Although she is inclined to die on Earth with everybody else, Nahautu consents to being transported to the Ring, where she lives out the rest of her days on a strip of land a quarter of a mile wide. Over the years as her grief dulls, she becomes accustomed to the place; comfortable, even. "But then I look up," says Nahautu, ending the story. What is it she sees? The infinite darkness of space? A high-tech astrodome? The toroidal arches of an O'Neill cylinder? Whatever's above her, it's not a sky.

The story's outcome is devastating, of course, but its proposal is a compelling one. What if we were to take O'Neill and Bezos seriously when they declare that the expansion into space will allow Earth to heal? What if we held them to it and sent anyone who wants endless energy, same-day shipping, and single-use plastics out to the asteroid belt, while the rest of us learn how to live within the quieter limits of Earth?

While we're creating worlds otherwise, what if we didn't even exile the 1-Clickers or sacrifice the asteroid belt? What if the lamas and rabbis and Franciscans and shamans could find some way to teach us all to find "another dimension of living life" wherever it is we are? Could that be the sort of Change we'd need to approach Earth *and* space differently?

Listening

In the fall of 2020, I was teaching a seminar called "Colonizing Space" to eighteen brilliant, scared, and exhausted students on

Zoom. We were reading most of the stuff I've been writing about here: the history of colonialism, the biblical legacy, imperial theology, astronautic policy, space law, Indigenous philosophy, and Afrofuturist worlding. I had patched the syllabus together in a mood of utter loneliness, wondering whether these connections would matter, or even make sense, to anyone else.

The students saw it immediately: the secular messiahs, the hypocritical humanisms, the effort to save the world with the same operations that have wrecked it. But we were just some liberal arts squares, staring at each other in literal squares. Out in the real world, a few courageous astrophysicists like Lucianne Walkowicz, Chanda Prescod-Weinstein, Erika Nesvold, and Parvathy Prem were giving TED talks, hosting "un-conferences," airing podcasts, and drumming up Twitterstorms to call attention to the growing danger of NewSpace, but not many people outside the space sciences were paying attention.

From my perspective, this anticolonial spacewave finally hit the public shores in October 2020, with the publication of a white paper written by the Equity, Diversity, and Inclusion Working Group of the Planetary Science and Astrobiology Decadal Survey for 2023–2032 (which makes recommendations to NASA, the National Science Foundation, and other government agencies). It's not that the American electorate was in the habit of reading decadal surveys and suddenly woke up with this one to the problem of space justice. It's that this publication—or perhaps the growing alliance that produced it—finally caught the attention of a critical mass of journalists, who started flooding the *Atlantic*, the *New Yorker*, the *New Republic*, *Slate*, *Vox*, and even *Space.com* with articles about the dangerous legacy of colonialism in space. Back at the University of Zoomlandia, I ditched whatever I'd had on my syllabus that day so we could read the white paper.

In this admittedly dense report, NASA communications specialist Frank Tavares and a host of coauthors argue that the capitalist-colonial approach to space is profoundly misguided, even dangerous. They remind the government agencies that the commodification of celestial bodies will pollute the solar-systemic environment, create disastrous working conditions in space, and exacerbate inequalities on Earth. They warn them that unchecked resource extraction might change these worlds forever, causing aesthetic offense to some humans and religious offense to others. They expose the chauvinism of public and private astro-utopias alike, whose imagined communities are limited to wealthy, young, able-bodied, heteroreproductive, and overwhelmingly white bodies. They decry the lack of regulation of private space actors, whom the spacefaring governments frankly seem to fear. And they criticize the Outer Space Treaty as "non-binding, outdated," and inadequate to "the realities of today's space industry."[57]

With all these problems in mind, the authors make a single, modest recommendation. They ask that national and international Planetary Protection agencies collect "community input" regarding their current priorities in space.[58] In other words, the coauthors advise, these agencies should tell the people what they're up to (the lunar outposts, the lunar mines, the asteroid mines, the Marshot), *and ask them what they think*. This last part will clearly be the hardest. NASA has created charming, even thrilling animated videos to explain the Artemis mission, but there are no opportunities for the public to ask questions about the strategically glossed-over details, much less to raise concerns about them, much less to raise objections.

"Input," the authors suggest. Not just output, not just infomercials, but input! The thoughts of us nonspecialists who do, after all, have our own kind of wisdom. We may not be rocket scientists, but we know about the system they're relying on. We know what the energy industry has done to our water, what the aeronautics industry has done to our skies, and what the space industry has done to our mountains. We know that transnational corporations don't value the health of their workers and that a thriving titanium economy does nothing to feed hungry people, give poor folks a living wage, or give anybody health care. We know that the only thing likely to trickle down from space is more pollution.

At the same time, we know other ways to go about things. Some of us can navigate by the stars; others can restore the balance of ecosystems; others can study and even work the land without destroying it. We know the values we try to teach our kids—like sharing, gentleness, listening, only taking what you need, and cleaning up after yourself—before their obligations to the market force them into the same awful compromises their parents have made. Are the space agencies ready to hear from the rest of us? Are they ready to hear that there are ancestors in the spaceways, temples in the mountains, creators in the planets? And are they ready to "adjust [their] practices and plans"[59] in the event that "community input" might ask them, for example, not to frack the Moon?

For Tavares and his colleagues, such adjustment would not mean shutting down the space program or stalling the pursuit of knowledge. Rather, it would mean learning other ways to live in relation to cosmic land and other ways to do science. As they propose,

> An alternative approach [to exploitation and commodification] can be found in Indigenous knowledge, which is inherently interdisciplinary, multigenerational, and expressed through sustainable practices. . . . Science in such a framework is not something done "on" or "to" land, but is created in relationship to a place and with deep intentionality and respect.[60]

Relationship, intentionality, and respect. These are some of the hallmarks of "traditional ecological knowledge" (TEK) which also values caretaking and listening.

As Tlingit anthropologist Kyle Wark explains, TEK's theoretical models begin with an account of the relationships between the human and nonhuman beings in any given environment (such beings might be siblings, reincarnated ancestors, etc.). These models then reflect on the mutual responsibilities between the environment's interdependent beings. "Humans are an integral component of the land," Wark writes; "we provide for it even as it provides for us."[61]

Perhaps most counterintuitively for Western institutions, TEK listens to the more-than-human world in order to learn from it. Plants know how to make food from sunlight. Mushrooms know how to send messages between trees. Geese know how to fly from Canada to North Carolina without a GPS, and my six-pound cat once managed to keep himself alive outside for three months without human-made shelter or factory-farmed food. (I'd have made it about a day and a half.) For these reasons, as Potawatomi biologist Robin Wall Kimmerer explains, "In Native ways of knowing, human people are often referred to as 'the younger brothers of Creation.' We say that humans have the least experience with how to live and thus the most to learn—we must look to our teachers among the other species for guidance."[62] What if our approach to space were guided by kinship, caretaking, and listening rather than conquest, war making, and profit?

Mars Society president Bob Zubrin isn't having any of it. In a *National Review* opinion piece titled "Wokeists Assault Space Exploration," Zubrin ridicules the white paper's authors for defending the rights of "dead rocks" at the expense of human progress.[63] For Zubrin, it seems, there are only two options: either extract, commercialize, and terraform, or shut down the space program. Since the "wokeists" are opposed to the first, they're clearly in favor of the second.

But to put it crudely, that's simply not true. This group of astrophysicists and social scientists unanimously adores space and wants to learn as much as they can about it. Some of them would even like to go there someday; they just don't want to wreck it in the process. So the authors aren't asking that we shut down space exploration; they're asking that we do it differently. With more respect, caretaking, and listening.

To Zubrin's ears, such value sound like absolute nonsense. The paper, he fumes, is useless on scientific grounds and amounts to nothing more than "ancient pantheistic mysticism and postmodern social thought."

It's always hard to know what someone is talking about when

they resort to name-calling. Rather than making an argument, name-callers use nasty words to dismiss a position automatically. As if it's not even worth considering. But let me take a moment to define Zubrin's terms.

By "postmodern social thought," I think Zubrin is referring to the authors' embrace of perspectivism. The kind that acknowledges that one person's "dead rock" is another person's sacred mountain and that refuses to belittle the second perspective as inferior to the first. When Zubrin dismisses the paper as "ancient pantheistic mysticism," he is engaging in precisely this sort of belittlement. Any form of knowledge that can be described as "ancient" is for Zubrin outdated and irrelevant. "Pantheism" is the idea that God is the natural world itself: clearly ridiculous. And "mysticism," or the unification of the individual spirit with God or the universe itself, is a major waste of time. "As a methodology for understanding the natural world," Zubrin claims, "mysticism has been displaced for some time by Western rationalism."

And he's not wrong! In the West, at least, a kind of thinking that calls itself rational has usurped traditional spiritual practices and religious knowing. At the same time, as we've seen throughout our journey, this rationalism preserves and even amplifies the most destructive elements of religious teachings, now divorced from their devotional contexts and masquerading as universal truth. Meanwhile, our Indigenous teachers, Saturnalian musicians, Black counterutopians, scientific heretics, and the leaders of the three largest Christian denominations cry out that we've simply got to stop trying to own the whole universe.

Who knows—maybe ancient pantheistic mysticism is exactly what we need.

Revolt of the Pantheists

Every part of this soil is sacred in the estimation of my people.
CHIEF SEATTLE

Four months into the 2020 shutdown, I was driving a three-year-old Elijah to his grandparents so we could get out of the damn house. (We weren't supposed to be seeing the grandparents, of course, but with an infant, a toddler, and two working parents, we didn't have much of a choice.) On the radio, Madonna was singing. I was most likely singing along. "I close my eyes / Oh God, I think I'm falling."

"God?" Elijah chirped from his car seat. "Who's God?"

A crowd of thoughts clamored for my attention. Had I really not told him about God? Had I not even used the word? How could I explain the historically shifting, culturally differential concept of God to a three-year-old? Should I tell him about the God of this song? The God of the Bible? *Which part of the Bible?*

But without listening to any of these questions, I found myself answering—really, the moment he asked—"God is all the stuff that makes the world."

"Oh," said Elijah. "I don't *like* God."

I burst out laughing. "You don't have much of a choice!"

"Ohhhhkayyyy," he relented. "I'll like God."

I had just written a book about pantheism, the idea that the world itself is God.[1] It's a teaching that springs up throughout Eastern, Indigenous, and Western traditions alike, and it's a teaching that Western philosophers tend to *hate*. With hilarious regularity, they call the idea juvenile, girly, primitive, and incoherent.

In Western philosophy and theology, "God" has traditionally meant a single, disembodied, humanoid, all-powerful intelligence outside the world. Monotheists will say that this God transcends gender while continuing to refer to God as "he." Meanwhile, the Western "world" is all the things its God is not: multiple, material, animal, vegetable, mineral, limited, and so forth. And although the world also surely transcends gender, Westerners tend nevertheless to associate it with femininity, calling Earth "mother" and the universe a "matrix" (that's just the same word in Latin). So the equation of "God" and "world" drives Western philosophers nuts. The religious ones say it's an insult to God to call him "just" the world, while the atheists say it's an insult to their intelligence to dress up the world by calling it God.

Meanwhile, many non-Western philosophers shrug their shoulders. They may disagree over the oneness or the manyness of God, or they may prefer the language of spirits and persons over the language of God, but the idea *that the universe itself is the source of creation, destruction, and renewal*—there's nothing particularly objectionable about that from most non-Western perspectives.

I've even found that individual Westerners aren't troubled by the idea—as long as they're not too conservative or rabidly atheistic and don't hold a PhD in philosophy or religion. Whenever I talk about pantheism to book clubs and church groups, stragglers of all ages, races, and genders will stay behind to confess that they've always kind of considered themselves to be pantheists. And although I'm always grateful for these expressions of intellectual solidarity, I tend to refrain from responding with any sort of personal identification. When people ask if I'm a pantheist, I'll say something like, "I'm not sure, but I think it's a powerful idea." Powerful because, as I've suggested throughout our journey with the astrotopians, we might be more inclined to respect and care for the world, its creatures, and its formations if they were sacred. But do I *believe* it? "Oh, I have no idea," I'll say; "it's the behavior that matters, not the belief."

And then out of nowhere, a three-year-old asks me who God is. His baby brother Ezra has kept me up all night for months, so I've got no energy to summon my usual qualifications and prevarications. They wouldn't go over well, anyway. Elijah wants an answer so I make it as simple as possible for him and as bearable as possible for me.

I don't go for the old man in the sky, nor do I tell him God is a character people made up, nor do I call God a "great force" behind the universe. Rather, I go for the little things. When I say "all the stuff that makes the world," I mean the microbes, minerals, electricities, magnetisms, attractions, repulsions, animals, and vegetables that create, sustain, destroy, and recycle everything that is. As far as I can see, these microcreators don't add up to a massive One, but they're not single actors, either. Rather, each thing works in relation to a whole slew of other things to make, unmake, and remake the world. And making, unmaking, and remaking the world is the work of whatever we think of as god(s). So that's what "God" means for this sort of pantheism: the work and unwork of creation in the hands of creation itself.

I'll stand by my usual tactics here and say that it doesn't matter whether this sort of pantheism, or any sort of pantheism, is "true." What matters is the way any given mythology prompts us to interact with the world we're part of—the world each of our actions helps to make and unmake. And frankly, some mythologies prompt us to act better than others.

As we've seen throughout this book, one of the most dangerous legacies of imperial Christianity is humanity's "dominion" over the rest of creation. Thanks to a particularly toxic reading of the first few chapters of Genesis, imperial Christianity leaves us with a starkly hierarchical cosmos in which Christians are superior to non-Christians, men are superior to women, humans are superior to animals, animals are superior to vegetables, and rocks are just rocks. Thinking ourselves separate from the rest of creation, Western humans have taken everything we can wrest from Earth—as if we're not part of the Earth we're destroying. As if we're not part of Earth's self-creation. But as we've also seen, there are other stories.

One of my favorites is the Cheyenne origin myth, whose creator needs help to make the world. Maheo the All-Spirit creates water, light, water people, and sky people, but then needs those creations to join in his work. Following the request of a weary Loon with nowhere to rest, Maheo and the water people together make dry land. They also make land people, including humanity.

As Pueblo scholar Paula Gunn Allen explains, this story presents a God of limited power, instilling values of listening and cooperation rather than unilateral declarations.[2] It describes human beings as the indebted products of animal cocreators rather than the rulers of creation. And this myth is just one example of the vast stores of Indigenous knowledge that Western rationality still thinks it's overcome. Just one of countless examples of different ways to live. As Jemisin reminds us, *there are other options*.

*   *   *

Ursula Le Guin opens her short story "Newton's Sleep" on a periapocalyptic Earth, the kind of scene to which she's helped us grow distressingly accustomed. War, climate disaster, and a "fungal plague" have made the planet uninhabitable, so some people move to a "dometown" on Earth while others try to find a way to leave it. Some of these others follow an aging prophet who has taught them to build a fully rational society on a rotating off-world cylinder, an O'Neill- or Bezos-type messiah who has grown too old to meet his own emigration criteria. His society is called Special Earth Satellite, or SPES for short (a word that sounds like "space" when you say it out loud and that means "hope" in Latin).

The story centers on the family of Isaac ("Ike") Rose, one of the community's leaders and a fervent devotee of Reason. Through Ike's internal monologues, we learn about the rigorous entrance exams for aspirational colonists, which have selected a society of eight hundred people whose average IQ is 165. Of these superior humans, almost every woman is of childbearing age. Everyone speaks English. There are seventeen Jews including Ike and his family. There are a handful of Asians, and no one is Black. "In a closed community of only eight hundred," Ike reasons, "every single person must be fit, not only genetically, but intellectually. And after the breakdown of public schooling during the Refederation, blacks just didn't get the training. . . . They were wonderful people, of course, but that wasn't enough."[3]

Not only are there no old people, very few people of color, and no disabled people (apart from Ike's daughter Esther, whose failing eyesight he intends to fix as soon as she's of age), but there are no creatures other than humans. No dogs, cats, insects, or even house plants, any of which could carry a virus that would wipe out the whole colony. One imagines the community has found some way to eat, but the food is probably manufactured in a sterile lab and freeze dried. In short, SPES is a gleaming construction of vinyl, plastic, and steel, free from all dirt and germs and maintaining a bright, balmy temperature all year round for its perfectly healthy, ingenious, emotionally stable pioneers.

The colony retains a few thin ties to their mother planet: "holovid" monitors that allow them to tune in to terrestrial events, a few classes in earthly history and science, and "landscapes" projected on the walls of their apartments that show mountains, gardens, oceans, and skies (the Rose family lives in "Vermont"). The least sentimental of the colony's leaders, Ike has long worried that these remnants will impede the colony's full enlightenment by tying them to a primitive world. So it's Earth he blames when his wife Susan reveals the escalating anti-Semitism among some of his colleagues. "My God!" he says (the interjection is strictly rhetorical; Ike is a staunch atheist). "We can keep out every virus, every bacterium, every spore, but this—this gets in? . . . I tell you, Susan, I think the monitors should be closed. Everything these children see and hear from earth is a lesson in violence, bigotry, superstition."[4]

But the more the colonists try to forget Earth, the more insistently it returns. When a committee suggests that the teachers cancel their geology lessons (who needs to know about rocks on SPES?), members of the colony start encountering things that aren't supposed to be there. First, some children see an old woman with white eyes, "burned all over."[5] Just as the men are dismissing their offspring as "hysterical little girls" à la Salem, Massachusetts, their wives see a group of Black people enter a room and then leave it again.[6] The men dismiss these escalating visions as a "mass hallucination" until they themselves start seeing scores of humans suffering from disease, washing out their laundry, and wearing animal skins. And even as goldfish come out of the bathroom tap, bison and wild horses run the pristine halls, and whales swim in the artificial seas, the obstinately rational Ike Rosen can't see any of it until he finally trips over a rock.[7]

As the visions multiply, different characters offer different interpretations of what might be going on. One calls the apparitions a visual expression of the community's guilt—presumably the guilt of having left the suffering Earth behind.[8] Another calls the apparitions "ghosts," the restless spirits of a wronged creaturekind. "It's going backwards, Dad,"[9] says Ike's son Noah (who like his whole family is named after a covenanted character from the Hebrew Bible). First, the colonists saw the humans to whom they'd judged themselves superior: old people, sick people, Black people. Then animals, then plants, and then finally Ike's rock, "pocked and cracked" and covered in "yellowish lichen."[10] It's the whole Western hierarchy of being, invading the overlords' gleaming utopia.

As Ike's wife Susan sees it, however, it's not an invasion at all. It's just an outward expression of everything the colonists are. "Humans," after all, are barely humans at all. We're 90 percent bacteria. We're animals made of vegetables and minerals and sunlight and starstuff. We're not just on Earth, we're *of* it, along with everything we think of ourselves as ruling. "How did we, how could we have thought we could just leave?" Susan asks. "Who do we think we are? All it is, is *we brought ourselves with us*. . . . The horses and the whales and the old women and the sick babies. They're just us, we're them, they're here."[11]

*   *   *

Should we explore outer space? Yes, if we can find a way to study it without doing further damage to its ecology and our own and without escalating human violence. Yes, if we can rein in private interest enough to privilege knowledge over profit and cooperation over competition. Should we try to live there? I'm honestly not sure. But either way, we need to stop pretending that escaping Earth is going to solve our problems. Because as Le Guin's SPES-niks learned, we'll bring them all along with us one way or another.

And it's precisely this bringing-it-all-along-with-us that might actually help. If the space agencies and corporations would actually take the time to listen to the "pantheistic mysticism" of Indigenous, Afrofuturist, feminist, and environmentalist communities—along with the more transcendent mysticisms of their monotheistic counterparts—they might realize that the fate of "humanity" cannot be separated from the fate of the rest of our Earth. That rich folks can't just leave the rest of us to weather the apocalypse, because the rest of us will return to disrupt their pristine space malls with our dragon clouds, companion species, angry gods, and sacred rocks. All those things that might actually carry us through the apocalypse. All the stuff that makes the world.

Acknowledgments

This book was funded by a fellowship at Wesleyan University's College of the Environment, directed by Barry Chernoff. It would have been frankly unwritable without the help of my cohort-mates David Grinspoon, Martha Gilmore, and Victoria Smolkin, along with COE affiliates Helen Poulos, Antonio Machado-Allison, and Alison Santos. And it would have been less of a pleasure to puzzle over without the students in my 2020 and 2021 "Colonizing Space" classes.

For orienting me to the small but energetic world of astrophysics and social justice, I am indebted to Meredith Hughes, Seth Redfield, Lucianne Walkowicz, Chanda Prescod-Weinstein, Ingrid LaFleur, Frank Tavares, and Parvathy Prem. For their help in navigating the dizzying field of space law, I am grateful to Jon Kravis, Brendan Cohen, Michelle Maloney, and especially Timiebi Aganaba. For consultations on Indigenous ontologies, mythologies, and politics, I have relied throughout the book on the invaluable assistance of J. Kēhaulani Kauanui (Kanaka Maoli). I am also thankful to Alvin Donel Harvey (Navajo Nation).

In an extraordinary senior thesis on "lunar fantasies and earthly supremacy," my former student Helen Handelman first got me interested in religion and the Apollo legacy. My cousin Patrick Guariglia tuned me in to what the tech guys were up to in outer space. Scott Mendel helped shape the book's early outline. Lori Gruen counseled me on the logistics of trade publishing. William

Robert helped edit the proposal. Katie Lofton, Jeff Kripal, Steve LaRue, and an anonymous reader made the manuscript better. And Kyle Wagner has guided the project with encouragement and care. My administrative and librarian colleagues Sheri Dursin, Kendall Hobbs, and Kate Wolfe have found and scanned more sources than I had any right to request. Francesca Baird and Marijane Ceruti performed their digital wizardry on the image files. Gabriel Urbina helped unravel some sonic puzzles along with my brother Kenan Rubenstein, who also dug me out of some nasty conceptual ditches.

The tireless, unmatched Winfield Goodwin has gone through the entire manuscript with a fine-toothed cursor. Uli Plass has read each chapter and offered invaluable insights into the utopian workings of capital. The book could not have found a more generous test reader than my mother, Veronica Warren, a more *critical* critical respondent than Simone Kotva, or a more capable final cheerleader than Catherine Keller. And I have relied throughout the writing process on the insight and forbearance of my partner Sheeja Thomas, along with our two kids, whose names I've changed but who know who they are. May they grow up knowing *where* they are, and what worlds we might still make.

Notes

Preface

1.   Keolu and Prescod-Weinstein, "Fight for Mauna Kea."

Introduction

1.   Sun Ra, "We Hold This Myth to Be Potential," in Abraham, *Sun Ra: Collected Works*, 1:210.
2.   Elon Musk, Twitter post, March 21, 2021, https://twitter.com/elonmusk/status/1373507545315172357.
3.   Russel and Vinsel, "Whitey on Mars."
4.   Cuthbertson, "Elon Musk's SpaceX Will 'Make Its Own Laws on Mars.'"
5.   Khatchadourian, "Elusive Peril of Space Junk."
6.   Tavares et. al., "Ethical Exploration."

Chapter One

1.   Davenport, *The Space Barons*, 42.
2.   Musk, "Making Humans a Multi-Planetary Species."
3.   Davenport, *The Space Barons* 26.
4.   Cao, "Jeff Bezos Thinks He's Winning the 'Billionaire Space Race.'"
5.   Bezos, cited in Davenport, *The Space Barons*, 147.
6.   Foer, "Jeff Bezos's Master Plan."
7.   "Kamala," Fandom, Memory Alpha, https://memory-alpha.fandom.com/wiki/Kamala.
8.   Jacobs, "Inside the Ultra-elite Explorers Club"; Davenport, *The Space Barons*, 197.
9.   Davenport, *The Space Barons*, 197.
10.   Musk, "Making Humans a Multi-Planetary Species."
11.   Musk, 46.
12.   Musk, 47.
13.   Musk, cited in Beers, "Selling the American Space Dream."
14.   Musk, "Making Humans a Multi-Planetary Species," 56.

15. Hore-Thorburn, "Trust Elon Musk to Make Going to Space Sound Shit."
16. Musk, "Making Humans a Multi-Planetary Species," 46.
17. In Davenport, *The Space Barons*, 244.
18. Musk, "Making Humans a Multi-Planetary Species," 46.
19. Zubrin and Wagner, *The Case for Mars*, 268.
20. Kaku, *The Future of Humanity*, 88.
21. "Elon Musk Will Need More Than 10,000 to Nuke Mars," TASS, https://tass.com/science/1155417.
22. Elon Musk, Twitter post, May 17, 2020, https://twitter.com/elonmusk/status/1262076013841805312.
23. Raz, "Lucianne Walkowicz."
24. Andersen, "Exodus."
25. "He brings salves and balm with him, no doubt; but before he can act as a physician he first has to wound; when he then stills the pain of the wound *he at the same time infects the wound*." Nietzsche, *On the Genealogy of Morals*, 3.15.
26. Vance, *Elon Musk*, 23-44.
27. Russell and Vinsel, "Whitey on Mars."
28. Bezos, cited in Davenport, *The Space Barons*, 259.
29. Sagan, *Pale Blue Dot*, 7.
30. John Carlos, Twitter post, February 14, 2021, https://twitter.com/JohnCarlos/status/1361093240280129541.
31. Bezos, cited in Davenport, *The Space Barons*, 258.
32. Bezos, cited in Davenport, 259.
33. Davenport, 260.
34. This dictum is usually attributed to Frederic Jameson, but some people attribute it to Slavoj Žižek.
35. Bezos, "Going to Space to Benefit Earth."
36. Bezos, "Going to Space."
37. "Blue Moon," Blue Origin, https://www.blueorigin.com/blue-moon/.
38. Roberge, "Elon Musk and Tesla."
39. Foer, "Jeff Bezos's Master Plan."
40. Roulette. "Elon Musk's Shot at Amazon."
41. Kaku, *The Future of Humanity*, 11.
42. H.R.2262—U.S. Commercial Space Launch Competitiveness Act, https://www.congress.gov/bill/114th-congress/house-bill/2262/text.
43. Pence, "Address."
44. See, for example, "Believes the Whole Book," *Chicago Chronicle*, April 2, 1897, https://www.newspapers.com/clip/76063969/moody-the-old-book. With thanks to Peter Manseau.
45. Psalm 139:10 (NRSV).
46. Armus, "Trump's 'Manifest Destiny' in Space."

Chapter Two

1. All biblical passages come from the New Revised Standard Version.
2. Prior, "The Right to Expel," 9.
3. Prior, *The Bible and Colonialism*, 177.

4.   Nietzsche, *On the Genealogy of Morals*, 3.24.

5.   Nietzsche.

6.   Nietzsche, 3.25.

7.   Elon Musk, Twitter post, June 12, 2021, https://twitter.com/elonmusk/status/1403921214234447890.

8.   White, "Historical Roots of Our Ecologic Crisis," 1204.

9.   White, 1203.

10. White, 1205.

11. Brown, "Pagan."

12. For critical treatments of the history, abuses, and potential transformations of this term, see Bird-David, "'Animism' Revisited"; Harvey, *Handbook of Contemporary Animism*.

13. White, "Historical Roots of Our Ecologic Crisis," 1205.

14. White, 1205.

15. See, for example, Erickson, "'I Worship Jesus, Not Mother Earth.'"

16. An illustrative range of these theologies, both within and beyond the Christian fold, can be found in Kearns and Keller, *Ecospirit*.

17. For a live collection of denominational statements on climate change, see the Yale Forum on Religion and Ecology, https://fore.yale.edu/Climate -Change/Climate-Change-Statements-World-Religions/Christianity -Protestant-Denominations-and. On the annual Blessing of the Animals at New York's Episcopal Cathedral of St. John the Divine, see Megan Roberts, "The Beastly Blessing: St. Francis Day at Manhattan's St. John the Devine," *Atlas Obscura*, October 9, 2013, https://www.atlasobscura.com/articles/a -beastly-blessing-st-francis-day-at-manhattan-s-cathedral-church-of-st -john-the-divine. Pope Francis describes the ministry of his namesake as follows: "whenever he would gaze at the sun, the moon or the smallest of animals, he burst into song, drawing all other creatures into his praise. He communed with all creation, even preaching to the flowers, inviting them to 'praise the Lord, just as if they were endowed with reason.'" Francis, "Laudato Si,'" ¶11.

18. Francis, ¶2.

19. Ecumenical Patriarch Bartholomew, Pope Francis, and the Archbishop of Canterbury, "A Joint Message for the Protection of Creation."

20. Ecumenical Patriarch Bartholomew, cited in Francis, "Laudato Si,'" ¶8; Evangelical Lutheran Church in America, "Caring for Creation." Francis, "Laudato Si,'" ¶67.

21. As Pope Francis argues, "Where profits alone count, there can be no thinking about the rhythms of nature, its phases of decay and regeneration, or the complexity of ecosystems which may be gravely upset by human intervention. Moreover, biodiversity is considered at most a deposit of economic resources available for exploitation, with no serious thought for the real value of things, their significance for persons and cultures, or the concerns and needs of the poor" Francis, "Laudato Si,'" ¶190.

22. Friedman, *Who Wrote the Bible?*, 50–51.

23. The Psalms reaffirm this order by exclaiming, "You have made [humans] a little lower than God, and crowned them with glory and honor. You have given them dominion over the works of your hands; you have put all things

under their feet, all sheep and oxen, and also the beasts of the field, and birds of the air, and fish of the sea, whatever passes along the paths of the seas" (Psalm 8:5–8).

24. See, for example, 1 Samuel 15:2–3 and Psalm 8:5–8.

25. See Prior, "Confronting the Bible's Ethnic Cleansing in Palestine," 11–12; *The Bible and Colonialism*, 290; Langston, "'A Running Thread of Ideals.'"

26. Warrior, "Canaanites, Cowboys, and Indians," 22; cf. Salaita, *The Holy Land in Transit*; Prior, *The Bible and Colonialism*, 39.

Chapter Three

1. Stone, *1776*, 74–75. For a historical account of Franklin's comparison of the eagle and the turkey, see Stamp, "American Myths."

2. "For my own part I wish the Bald Eagle had not been chosen the Representative of our Country. He is a Bird of bad moral Character. He does not get his Living honestly. You may have seen him perched on some dead Tree near the River, where, too lazy to fish for himself, he watches the Labour of the Fishing Hawk; and when that diligent Bird has at length taken a Fish, and is bearing it to his Nest for the Support of his Mate and young Ones, the Bald Eagle pursues him and takes it from him." Franklin in Stamp, "American Myths."

3. John Adams, letter to Abigail Adams, August 14, 1776, cited in Patterson, *The Eagle and the Shield*, 16–17.

4. Benjamin Franklin, undated note, August 1776, in Patterson, *The Eagle and the Shield*, 14.

5. For an exhaustive sourcebook on and analysis of this metaphor, see Cherry, *God's New Israel*. Steven Salaita points out that "there are more than twenty towns in the United States named Canaan or New Canaan, and several Palestines . . . in Arkansas, Illinois, Texas, and West Virginia." Salaita, *The Holy Land in Transit*, 13.

6. Cherry, *God's New Israel*, 1.

7. Cited in Langston, "'A Running Thread of Ideals,'" 239.

8. See the accusations leveled by Tomás Ortiz cited in Prior, *The Bible and Colonialism*, 181.

9. See Mather, *Soldiers Counselled and Comforted*, 32; Wigglesworth, "God's Controversy with New England (1662)," line 7; and Whitaker, "Good Newes from Virginia (1613)," 32–33.

10. See Edwards, "Latter-Day Glory," 55.

11. Sepulveda, "Democrates Alter."

12. Lunar Embassy, "Buy Land on the Moon," https://lunarembassy.com/product/buy-land-on-the-moon/.

13. Lunar Embassy, "About Our Founder Dennis Hope," https://lunarembassy.com/who-owns-the-moon-dennis-hope/.

14. "The Man Who Sells The Moon," Op-Docs, *New York Times*, March 11, 2013, https://www.youtube.com/watch?v=Bs6rCxU_IHY.

15. Alexander VI, *Inter caetera*.

16. Nicholas V, "*Romanus Pontifex*."

17. Alexander VI, *Inter caetera*.

18. Alexander VI, "emphasis added.

19. Council of Castile, "Requerimiento."
20. Council of Castile.
21. Cited in Wynter, "The Pope Must Have Been Drunk."
22. Cited in Bartels, "Should We Colonize Space."
23. Emmerich de Vattel set the theoretical stage for the codification of *terra nullius* by applying John Locke's labor theory of property to the case of Indigenous Americans. "Their unsettled unhabitation in those immense regions cannot be accounted a true and legal possession," he insisted; "and the people of Europe, too closely pent up at home, finding land of which the savages stood in no particular need, and of which they made no actual and constant use, were lawfully entitled to take possession of it, and settle it with colonies." Vattel, *The Law of Nations*, §209.
24. Bartolomé de las Casas provides a firsthand account of the unthinkably cruel exploitation and mass murder of Caribbean Natives in de Las Casas, *History of the Indies*.
25. For the details of these rituals, see Seed, *Ceremonies of Possession*.
26. Eliade, *Myth of the Eternal Return*, 9-10.
27. On the connection between the Christian God's purported creation out of nothing and the political doctrine of *terra nullius*, see Bauman, "*Creatio ex Nihilo*."
28. Franklin cited in Donaldson, "Joshua in America,", 274, emphasis original.
29. See Langston, "'A Running Thread of Ideals,'" 246-47.
30. Thomas Jefferson, Letter to William Short, August 4, 1820, in Healey, "Jefferson on Judaism."
31. These examples only scratch the surface of these fathers' ethical ambivalence. The purportedly egalitarian Jefferson owned over six hundred slaves during his lifetime, and Franklin objected as strenuously to racial miscegenation as he did to the preservation of North American forests. As he saw it, both practices contributed to the darkening of the globe and risked offending—wait for it—the superior beings on Mars and Venus, who could see the continent much more clearly now that its white folks had clear-cut the old growth. See Donaldson, "Joshua in America," 276.
32. Trump, "Remarks."
33. "Annexation," *Democratic Review* 17 (July, 1845), 5. On the contested authorship of this piece, see Howe, *What Hath God Wrought*, 703, emphasis added.
34. Cherry, *God's New Israel*, 117.
35. Both presidents cited in Finkelstein, *Image and Reality*, 91.
36. "Annexation," emphasis added.
37. Cherry, *God's New Israel*, 113.
38. Chief Seattle, "Oration (1854)," 135.

Chapter Four

1. Retold in Young, "'Pity the Indians of Outer Space,'" 273.
2. Two exemplary texts in this regard are Wilkins, *Discovery of a World in the Moone* and Fontenelle, *Conversations on the Plurality of Worlds*.
3. This allegedly American characteristic was most famously propounded in the late nineteenth century by Frederick Jackson Turner, who stated that "American social development has been continually beginning over again

on the frontier. This perennial rebirth . . . furnish[es] the forces dominating American character." Turner, "Significance of the Frontier." More recently, Elon Musk said of the US that it is "a distillation of the human spirit of exploration. . . . You couldn't ask for a group of people that are more interested in exploring the frontier." Cited in Davenport, *The Space Barons*, 117.

4. As Eisenhower's presidential science advisor testifies, the event left him feeling "psychologically vulnerable and technically surprised. . . . I recall one night looking into the heavens, seeing *Sputnik*, and feeling an awesome, poetic sense of wonder. This instinctive human response to astronomical phenomena that seem to transcend man's natural ken was certainly an element in the stunned surprise of most to the news of *Sputnik I*." Killian, *Sputnik, Scientists, and Eisenhower*, 2.

5. Arendt cited in Lazier, "Earthrise," 602n2.

6. "Orderly Formula," *Time*, October 28, 1957, 17–19.

7. Kennedy, "Special Message to the Congress."

8. Kennedy, "If the Soviets Control Space."

9. Johnson cited in Killian, *Sputnik, Scientists, and Eisenhower*, 9.

10. Eisenhower, "Statement by the President."

11. Kennedy, "If the Soviets Control Space."

12. Kennedy, "Special Message to the Congress."

13. "What Are We Waiting For?," *Collier's*, March 22, 1952, 23.

14. Wernher von Braun, "Crossing the Last Frontier," *Collier's*, March 22, 1952, 29.

15. *Holocaust Encyclopedia*, s.v. "Lebensraum," https://encyclopedia.ushmm .org/content/en/article/lebensraum. Although Hitler enacted this doctrine with unprecedented brutality, *Lebensraum* preceded the advent of National Socialism by generations. See Zimmerer, "Birth of the Ostland Out of the Spirit of Colonialism." With thanks to Yaniv Feller.

16. See Newell, "Strange Case of Dr. von Braun and Mr. Disney."

17. Newell, "Strange Case of Dr. von Braun and Mr. Disney."

18. CBS News, "Live TV Transmission from Apollo 8."

19. Cited in Potter, *The Earth Gazers*, 293.

20. See Smolkin, *A Sacred Space*, 87–94.

21. United Nations Office for Outer Space Affairs, Committee on the Peaceful Uses of Outer Space, "Treaty on Principles Governing the Activities of States."

22. O'Neill, *The High Frontier*, 7.

23. Zubrin and Wagner, *The Case for Mars*, 323–34.

24. Davenport, *The Space Barons*.

25. Bezos, "Open Letter to Administrator Nelson."

26. Elon Musk, Twitter post, April 26, 2021, https://twitter.com/elonmusk/ status/1386825367948644352.

27. Döpfner, "Jeff Bezos Reveals What It's Like to Build an Empire."

28. President's Science Advisory Committee, "Introduction to Outer Space."

29. *Pioneering the Space Frontier: The Report of the National Commission on Space*, n.d., https://www.nasa.gov/pdf/383341main_60%20-%2020090814.5.The %20Report%20of%20the%20National%20Commission%20on%20Space .pdf.

30. Reported in Thompson, "Monetizing the Final Frontier," and Haskins, "Racist Language of Space Exploration."

31. George Takei, Twitter post, January 24, 2020, https://twitter.com/GeorgeTakei/status/1220823578825887745.

32. George Takei, Twitter post, December 18, 2020, https://twitter.com/GeorgeTakei/status/1340090866237546496.

33. Reported in Welna, "Space Force Bible Blessing at National Cathedral Sparks Outrage."

34. With thanks to Robert Yelle.

35. United States Space Force commercial, May 6, 2020, https://www.youtube.com/watch?v=9ud7wgbBBnY.

36. *Space Force*, episode 1:3.

37. *Space Force*, episode 1:1.

38. United States Space Force commercial, October 28, 2020, https://www.youtube.com/watch?v=x619VW65l1Y

39. United States Space Force, *Spacepower*, iv.

40. United States Space Force, *Spacepower*, vi.

41. United States Space Force, *Spacepower*, 53.

42. United States Space Force, *Spacepower*, v.

43. United States Space Force, *Spacepower*, xiii.

Chapter Five

1. CBS News, July 20, 1969, https://www.youtube.com/watch?v=gg5Ncc9GODY.

2. Willis Shapley, cited in Waxman, "Lots of People Have Theories about Neil Armstrong's 'One Small Step for Man' Quote."

3. See Gorman, "Cultural Landscape of Interplanetary Space," 100.

4. Tutton, "Socitechnical Imaginaries and Techno-Optimism."

5. United Nations Office for Outer Space Affairs, Committee on the Peaceful Uses of Outer Space, "Report of the Legal Subcommittee on Its Sixtieth Session."

6. See Foster and Goswami, "What China's Antarctic Behavior Tells Us"; United States Department of Defense, "Final Report on Organizational and Management Structure for the National Security Space Components."

7. United States Office of Space Commerce, "National Space Policy of the United States of America."

8. United States Department of Defense, "Final Report on Organizational and Management Structure for the National Security Space Components," 3.

9. United States Space Force, *Spacepower*, v.

10. United Nations Office for Outer Space Affairs, Committee on the Peaceful Uses of Outer Space, "Agreement Governing the Activities of States."

11. Cohen, "So You Want to Buy a Space Company?"

12. Dovey, "Mining the Moon."

13. Wall, "New Space Mining Legislation."

14. Executive Office of the President, "Encouraging International Support for the Recovery and Use of Space Resources."

15. Pace, "Space Development, Law, and Values."

16. United Nations Office for Outer Space Affairs, Committee on the Peaceful Uses of Outer Space, "Report of the Legal Subcommittee on Its Sixtieth Session."
17. NASA, "The Artemis Plan."
18. United States Office of Space Commerce, "National Space Policy of the United States of America."
19. United States Department of Defense, "Final Report on Organizational and Management Structure for the National Security Space Components of the Department of Defense."
20. President's Science Advisory Committee, "Introduction to Outer Space."
21. NASA, "Artemis Plan."
22. Andrews, "NASA Just Broke the 'Venus Curse.'"
23. NASA, "Artemis Plan," 58.
24. NASA, Artemis Plan, 60.
25. Quinn, "New Age of Space Law," 487.
26. AlinaUtrata, "Lost in Space."
27. Ayn Rand, *Atlas Shrugged*.
28. Charmaine Curtis Jacobs, "Letters to the Editor: Greta Thunberg's Powerful Rebuke of 'Fairy Tales of Eternal Economic Growth,'" *Los Angeles Times*, September 24, 2019, https://www.latimes.com/opinion/story/2019-09-24/greta-thunberg-climate-change-speech.
29. See Rubenstein, *Worlds without End*, 212.
30. Mohanta, "How Many Satellites Are Orbiting the Earth in 2021?"
31. See https://www.celestis.com/ and https://www.beyondburials.com.
32. Resnick, "Apollo Astronauts Left Their Poop on the Moon."
33. Millard, "The Only Bible on the Moon."
34. Khatchadourian, "Elusive Peril of Space Junk."
35. See https://astroscale.com/.
36. See Haroun et al., "Toward the Sustainability of Outer Space," 64.
37. Dovey, "Mining the Moon."
38. See http://astria.tacc.utexas.edu/AstriaGraph/.
39. Jah, "Acta non verba."
40. Sheets, "FCC Approves SpaceX Change to Its Starlink Network."
41. Ingraham, "Proliferation of Space Junk."
42. Lee, "'Silent Spring' Is Now Noisy Summer." My thanks to Ceridwen Dovey for pointing out this connection in her introductory comments to Annie Handmer, "MVA Public Forum on the Moon," Space Junk podcast, August 18, 2020, https://www.youtube.com/watch?v=8SB_ZwVgGOs.
43. Rawls et al., "Satellite Constellation Internet Affordability and Need."
44. Adebola and Adebola, "Roadmap for Integrated Space Applications in Africa"; Nakahado, "Should Space Be Part of a Development Strategy?"
45. LaFleur and Jah, "What Does the Afrofuture Say?"
46. Erwin, "Space Force Sees Need for Civilian Agency to Manage Congestion."

Chapter Six

1. Davenport, *The Space Barons*, 71.
2. Francis, "Laudato Si,'" ¶82.
3. Von Braun, "For Space Buffs."

4. See Day, "Paradigm Lost."
5. Von Braun, "Space Buffs."
6. Gorman, "Cultural Landscape of Interplanetary Space," 89.
7. Berry in Brand, *Space Colonies*, 36.
8. Kaku, *The Future of Humanity*, 55.
9. "Ten Things You Should Know about Bennu," NASA, October 16, 2020, https://www.nasa.gov/feature/goddard/2020/bennu-top-ten.
10. Kaku, *The Future of Humanity*, 58.
11. "Psyche," Jet Propulsion Laboratory, California Institute of Technology, https://www.jpl.nasa.gov/missions/psyche.
12. Klinger, *Rare Earth Frontiers*, 46.
13. Klinger, 7.
14. Klinger, 3.
15. Klinger, 4–5.
16. Andersen, "Exodus."
17. Tutton, "Socitechnical Imaginaries and Techno-Optimism," 11.
18. Zubrin, "Why We Humans Should Colonize Mars!," 308.
19. Zubrin, 306.
20. McKay, "Planetary Ecosynthesis on Mars," 245.
21. McKay, 264.
22. Margulis and West, "Gaia and the Colonization of Mars."
23. For an introduction to the Gaia hypothesis, see Lovelock, *Gaia*.
24. Even then, as Chris McKay points out, life on Mars would not last forever—not even until the death of our shared sun. Because Mars seems only to have one tectonic plate, it cannot recycle materials effectively. So even a flourishing microbial masterpiece would only last between 10 million and 100 million years (by comparison, life has existed on Earth for 4.5 billion years).
25. Zubrin and Wagner, *The Case for Mars*, 325.
26. Turner, "Significance of the Frontier."
27. Zubrin and Wagner, *The Case for Mars*, 325.
28. Stirone, "Mars Is a Hellhole."
29. Zubrin, "Why We Humans Should Colonize Mars!," 305.
30. Zubrin and Wagner, *The Case for Mars*, 334.
31. Zubrin and Wagner, *The Case for Mars*, 270–71.
32. Zubrin, "Why We Humans Should Colonize Mars!," 309, emphasis added.
33. Musk, cited in Andersen, "Exodus."
34. Fogg, "Ethical Dimensions of Space Settlement."
35. Walkowicz, "Let's Not Use Mars as a Backup Planet."
36. McKay, "Planetary Ecosynthesis on Mars"; Schwartz, "On the Moral Permissibility of Terraforming."
37. Margulis and West, "Gaia and the Colonization of Mars," 279.
38. Kaçar, "Do We Send the Goo?"
39. Sagan, 138.
40. Grinspoon, *Lonely Planets*, 97–114.
41. Steinem and Ride, "Sally Ride on the Future in Space."
42. Treviño, "The Cosmos is Not Finished."
43. Rolston, "Preservation of Natural Value in the Solar System," 147, emphasis in original.
44. Rolston, 173.

45. Rolston, 167.
46. See, for example, Schwartz, "On the Moral Permissibility of Terraforming," 14.
47. Sparrow, "The Ethics of Terraforming," 227.
48. Sparrow, 237.
49. Cohen and Tilman, "Biosphere 2 and Biodiversity," 1150.
50. Zach Rosenberg, "This Congressman Kept the U.S. and China from Exploring Space Together."
51. "Statement of Robert (Bob) Richards, Founder and CEO, Moon Express, Inc., before the United States House of Representatives' Committee on Science, Space and Technology Subcommittee on Space Hearing on Private Sector Lunar Exploration," September 7, 2017, https://science.house.gov/imo/media/doc/Richards%20Testimony.pdf.
52. Moon Treaty, Article 7.
53. Dovey, "Mining the Moon."
54. Kminek et al., "COSPAR Planetary Protection Policy," 13–14.
55. NASA, "The Artemis Accords," §9.
56. Savage cited in Dovey, "Mining the Moon."
57. Australian Earth Laws Alliance, "Declaration of the Rights of the Moon."
58. Emphasis in original.
59. Thomas Berry, "Rights of the Earth," 29.
60. Zubrin, "Wokeists Assault Space Exploration."
61. Gorman in Handmer, "MVA Public Forum on the Moon."
62. Gorman in Handmer.
63. Potter, *The Earth Gazers*, 321.
64. Gorman in Handmer, "MVA Public Forum on the Moon."
65. Sparrow, "The Ethics of Terraforming."
66. Maloney in Handmer, "MVA Public Forum on the Moon."
67. Bawaka Country, "Dukarr Lakarama."
68. Lampen, "Have TikTok Witches Actually 'Hexed the Moon'?"
69. As recounted in Young, "'Pity the Indians of Outer Space,'" 272.

Chapter Seven

1. Lewis-Kraus, "How the Pentagon Started Taking U.F.O.s Seriously."
2. See https://moonexpress.com/.
3. Thompson, "Monetizing the Final Frontier."
4. Robbins, "How Can Our Future Mars Colonies Be Free of Sexism and Racism?"
5. Utrata, "Lost in Space."
6. Wynter, "The Pope Must Have Been Drunk."
7. Bostrom cited in Torres, "Dangerous Ideas of Longtermism and Existential Risk
8. The reorientation of Christian theologies to "this-worldliness" is often attributed to the influence of Dietrich Bonhoeffer, who died in a concentration camp after having been part of a plot to kill Hitler. See Bonhoeffer, *Letters and Papers from Prison*.
9. See Torres, "Dangerous Ideas of Longtermism and Existential Risk."
10. Bostrom, "Are We Living in a Computer Simulation?"

11. Hanson, "How to Live in a Simulation."
12. Musk cited in Andersen, "Exodus."
13. Le Guin, "Ones Who Walk Away," 279.
14. Le Guin, 282.
15. Le Guin, 284.
16. Jemisin, "Ones Who Stay and Fight," 4.
17. Jemisin, 7, 9.
18. Jemisin, 11.
19. Jemisin, 11.
20. Brown, *Black Utopias*, 15–16.
21. Nietzsche, *On the Genealogy of Morals*, 3.25.
22. Ursula K. Le Guin, "Speech in Acceptance of the National Book Foundation Medal for Distinguished Contribution to American Letters," https://www .ursulakleguin.com/nbf-medal.
23. Cornum, "Creation Story Is a Spaceship."
24. Dillon, "Imagining Indigenous Futurisms."
25. Silko, *Ceremony*.
26. Cornum, "Creation Story is a Spaceship."
27. Eshun, "Further Considerations on Afrofuturism."
28. Dery, "Black to the Future."
29. See Wabuke, "Afrofuturism."
30. Ra cited in Szwed, *Space Is the Place*, 29.
31. Ra cited in Youngquist, *Pure Solar World*, 70.
32. Szwed, *Space Is the Place*, 295, 356.
33. Baraka, *This Planet Is Doomed*, 12.
34. Baraka, 34.
35. Sun Ra, "We Travel the Spaceways," Sun Ra and His Arkestra Greatest Hits, https://www.youtube.com/watch?v=oLn1JVsISho.
36. Ra, cited in Youngquist, *Pure Solar World*, 69–70.
37. Abraham, *Sun Ra: Collected Works*, 1:xxix.
38. Eshun, *More Brilliant Than the Sun*, 175.
39. Butler, *Parable of the Sower*, 195.
40. Butler, *Parable of the Talents*, 222.
41. Butler, 84.
42. Butler, 222.
43. Butler, 405.
44. Butler, 404.
45. Butler, 380.
46. Butler, 404–5.
47. Butler, 406.
48. Brown, *Black Utopias*, 107.
49. Butler, *Parable of the Talents*, 63–64.
50. Jemisin, "Cloud Dragon Skies," 119.
51. Jemisin, 118.
52. Jemisin, 118.
53. Jemisin, 118.
54. Jemisin, 120.
55. Jemisin, 119.
56. Jemisin, 119.

57. Tavares et al., "Ethical Exploration," 6.
58. Tavares et al., 1.
59. Tavares et al., 5.
60. Tavares et al., 5.
61. Wark, "Wis2dom," 103.
62. Kimmerer, *Braiding Sweetgrass*, 9.
63. Zubrin, "Wokeists Assault Space Exploration."

Conclusion

1. Rubenstein, *Pantheologies*.
2. Allen, *The Sacred Hoop*, 60.
3. Le Guin, "Newton's Sleep," 35.
4. Le Guin, 32.
5. Le Guin, 34.
6. Le Guin, 35, 42–43.
7. Le Guin, 53.
8. Le Guin, 45.
9. Le Guin, 49.
10. Le Guin, 53.
11. Le Guin, 51, emphasis added.

Bibliography

Abraham, Adam, ed. *Sun Ra: Collected Works*. Vol. 1, *Immeasurable Equation*. Chandler, AZ: Phaelos, 2005.

Adebola, Olufunke, and Simon Adebola. "Roadmap for Integrated Space Applications in Africa." *New Space* 9, no. 1 (2021): 12–18.

Aganaba, Timiebi. "Innovative Instruments for Space Governance." Center for International Governance Innovation, February 8, 2021. https://www.cigionline.org/articles/innovative-instruments-space-governance/.

Alexander VI. *Inter caetera*. May 4, 1493. *Papal Encyclicals Online*. https://www.papalencyclicals.net/category/alex06.

Allen, Paula Gunn. *The Sacred Hoop: Recovering the Feminine in American Indian Traditions*. Boston: Beacon Press, 1992.

Andersen, Ross. "Exodus." *Aeon*, September 30, 2014. https://aeon.co/essays/elon-musk-puts-his-case-for-a-multi-planet-civilisation.

Andrews, Robin George. "NASA Just Broke the 'Venus Curse': Here's What It Took." *Scientific American*, June 2, 2021. https://www.scientificamerican.com/article/nasa-just-broke-the-venus-curse-heres-what-it-took/.

"Annexation." *Democratic Review* 17 (July 1845): 5.

Armus, Teo. "Trump's 'Manifest Destiny' in Space Revives Old Phrase to Provocative Effect." *Washington Post*, February 5, 2020. https://www.washingtonpost.com/nation/2020/02/05/trumps-manifest-destiny-space-revives-old-phrase-provocative-effect/.

Australian Earth Laws Alliance. "Declaration of the Rights of the Moon." February 11, 2021. https://www.earthlaws.org.au/moon-declaration/.

Baraka, Amiri, ed. *This Planet Is Doomed: The Science Fiction Poetry of Sun Ra*. New York: Kicks Books, 2011.

Bartels, Meghan. "Should We Colonize Space? Some People Argue We Need to Decolonize It Instead." *Newsweek*, May 25, 2018. https://www.newsweek.com/should-we-colonize-space-some-people-argue-we-need-decolonize-it-instead-945130.

Bauman, Whitney. "*Creatio Ex Nihilo, Terra Nullius*, and the Erasure of Presence." In *Ecospirit: Religions and Philosophies for the Earth*, edited by Laurel Kearns and Catherine Keller, 353–72. New York: Fordham University Press, 2007.

Bawaka Country. "Dukarr Lakarama: Listening to Guwak, Talking Back to Space

Colonization." *Political Geography* 81 (August 2020). https://www.sciencedirect
.com/science/article/pii/S0962629818304086.

Beers, David. "Selling the American Space Dream: The Cosmic Delusions of Elon
Musk and Wernher Von Braun." *New Republic*, December 7, 2020. https://
newrepublic.com/article/160268/selling-american-space-dream.

Berry, Thomas. "Rights of the Earth." *Resurgence* 214 (September/October 2002):
28–29.

Bezos, Jeff. "Going to Space to Benefit Earth." Blue Origin, May 10, 2019. https://
www.youtube.com/watch?v=GQ98hGUe6FM&t=203s.

———. "Open Letter to Administrator Nelson." Blue Origin, July 26, 2021. https://
www.blueorigin.com/news/open-letter-to-administrator-nelson.

Bird-David, Nurit. "'Animism' Revisited: Personhood, Environment, and Rela-
tional Epistemology." In "Culture: A Second Chance?" Supplement, *Current
Anthropology* 40, no. S1 (February 1999): S67–S91.

Bonhoeffer, Dietrich. *Dietrich Bonhoeffer Works*. Vol. 8, *Letters and Papers from
Prison*. Translated by Isabel Best, Lisa E. Dahill, Reinhard Krauss, Nancy
Lukens, Barbara Rumscheidt, and Martin Rumscheidt. Edited by Eberhard
Bethge, Ernst Feil, Christian Gremmels, Wolfgang Huber, Hans Pfeifer,
Albrecht Schönherr, Heinz Eduard Tödt, and Ilse Tödt. Minneapolis, MN:
Augsburg Fortress Press, 2010.

Bostrom, Nick. "Are We Living in a Computer Simulation?" *Philosophical Quarterly*
53, no. 211 (April, 2003): 243–55.

Brand, Stewart, ed. *Space Colonies*. New York: Penguin, 1977.

Braun, Wernher von. "Crossing the Last Frontier." *Collier's*, March 22, 1952.

Brown, Jayna. *Black Utopias: Speculative Life and the Music of Other Worlds*. Durham, NC: Duke University Press, 2021.

Brown, Peter. "Pagan." In *Late Antiquity: A Guide to the Postclassical World*, edited
by G. W. Bowersock, P. R. L. Brown, and O. Grabar. Cambridge, MA: Harvard
University Press, 1999.

Butler, Octavia E. *Parable of the Sower*. New York: Grand Central Publishing, 2000.
First published 1993 by Four Walls, Eight Windows (New York).

———. *Parable of the Talents*. New York: Grand Central Publishing, 2000. First
published 1998 by Seven Stories Press (New York).

Cao, Sisi. "Jeff Bezos Thinks He's Winning the 'Billionaire Space Race.'" *Observer*,
February 25, 2019. https://observer.com/2019/02/amazon-jeff-bezos-blue
-origin-space-race/.

CBS News. "Live TV Transmission from Apollo 8." December 24, 1968. https://
www.youtube.com/watch?v=1aIfoG2PtHo&t=435s.

Cherry, Conrad, ed. *God's New Israel: Religious Interpretations of American Destiny*.
Rev. and expanded ed. Chapel Hill: University of North Carolina Press, 1998.

Chief Seattle. "Oration (1854)." In *God's New Israel: Religious Interpretations of
American Destiny*, edited by Conrad Cherry, 135–36. Chapel Hill: University of
North Carolina, 1998.

Cohen, Brendan. "So You Want to Buy a Space Company?" *International Institute of Space Law* 2 (2018), https://www.elevenjournals.com/tijdschrift/iisl/
2018/2%20Financing%20Space:%20Procurement,%20Competition%20and
%20Regulatory%20Approach/IISL_2018_061_002_004.

Cohen, Joel E., and David Tilman. "Biosphere 2 and Biodiversity: The Lessons So
Far." *Science* 274, no. 5290 (November 15, 1996): 1150–51.

Cornum, Lou. "The Creation Story Is a Spaceship: Indigenous Futurism and Decolonial Deep Space." Westar Institute, Fall 2019 National Meeting, November 22–26, San Diego, California. https://www.westarinstitute.org/fall-2019 -national-meeting/.

Council of Castile. "Requerimiento." 1510. National Humanities Center Resource Toolbox, *American Beginnings: The European Presence in North America, 1492– 1690*. https://nationalhumanitiescenter.org/pds/amerbegin/contact/text7/ requirement.pdf.

Country, Bawaka. "Dukarr Lakarama: Listening to Guwak, Talking Back to Space Colonization." *Political Geography* 81 (August, 2020). https://www .sciencedirect.com/science/article/pii/S0962629818304086.

Cuthbertson, Anthony. "Elon Musk's SpaceX Will 'Make Its Own Laws on Mars.'" *Independent*, October 28, 2020. https://www.independent.co.uk/life-style/ gadgets-and-tech/elon-musk-spacex-mars-laws-starlink-b1396023.html.

Davenport, Christian. *The Space Barons: Elon Musk, Jeff Bezos, and the Quest to Colonize the Cosmos*. New York: PublicAffairs, 2019.

Day, Dwayne A. "Paradigm Lost." *Space Policy* 11, no. 3 (1995): 153–59.

de Las Casas, Bartolomé. *History of the Indies (1561)*. New York: Harper and Row, 1971.

Dery, Mark. "Black to the Future: Interviews with Samuel R. Delany, Greg Tate, and Tricia Rose." In *Flame Wars: The Discourse of Cyberculture*, edited by Mark Dery, 179–222. Durham, NC: Duke University Press, 1994.

Dillon, Grace L. "Imagining Indigenous Futurisms." In *Walking the Clouds: An Anthology of Indigenous Science Fiction*, edited by Grace L. Dillon, 1–12. Tucson: University of Arizona Press, 2012.

Donaldson, Laura E. "Joshua in America: On Cowboys, Canaanites, and Indians." In *The Calling of the Nations: Exegesis, Ethnography, and Empire in a Biblical-Historic Present*, edited by Mark Vessey, Sharon V. Betcher, Robert A. Daum, and Harry O. Maier, 272–90. Toronto: University of Toronto Press, 2011.

Döpfner, Mathias. "Jeff Bezos Reveals What It's Like to Build an Empire." *Insider*, April 28, 2018. https://www.businessinsider.com/jeff-bezos-interview-axel -springer-ceo-amazon-trump-blue-origin-family-regulation-washington-post -2018-4.

Dovey, Ceridwen. "Mining the Moon." *The Monthly*, July 2019. https://www .themonthly.com.au/issue/2019/july/1561989600/ceridwen-dovey/mining -moon.

Ecumenical Patriarch Bartholomew, Pope Francis, and the Archbishop of Canterbury. "A Joint Message for the Protection of Creation." September 1, 2021, https://www.archbishopofcanterbury.org/sites/abc/files/2021-09/Joint %20Statement%20on%20the%20Environment.pdf.

Edwards, Jonathan. "The Latter-Day Glory Is Probably to Begin in America (1830)." In *God's New Israel: Religious Interpretations of American Destiny*, edited by Conrad Cherry, 54–58. Chapel Hill: University of North Carolina Press, 1998.

Eisenhower, Dwight D. "Statement by the President." In *NASA's Origins and the Dawn of the Space Age*. Washington, DC: NASA Historical Reference Collection, NASA History Office, NASA Headquarters, 1958.

Eliade, Mircea. *The Myth of the Eternal Return: Cosmos and History*. Translated by Willard R. Trask. Princeton, NJ: Princeton University Press, 2005.

Erickson, Jacob J. "'I Worship Jesus, Not Mother Earth': Exceptionalism and the Paris Withdrawal." *Religion Dispatches*, June 2, 2017. http://religiondispatches .org/i-worship-jesus-not-mother-earth-american-christian-exceptionalism -and-the-paris-withdrawal/.

Erwin, Sandra. "Space Force Sees Need for Civilian Agency to Manage Conges- tion." *Space News*, April 26, 2021.

Eshun, Kodwo. "Further Considerations on Afrofuturism." *Centennial Review* 3, no. 2 (Summer 2003): 287–302.

——. *More Brilliant Than the Sun: Adventures in Sonic Fiction*. London: Quartet Books, 1999.

Evangelical Lutheran Church in America. "Caring for Creation: Vision, Hope, and Justice." August 28, 1993. https://download.elca.org/ELCA%20Resource %20Repository/EnvironmentSS.pdf.

Executive Office of the President. "Encouraging International Support for the Recovery and Use of Space Resources." Executive order, April 6, 2020. https://trumpwhitehouse.archives.gov/presidential-actions/executive-order -encouraging-international-support-recovery-use-space-resources/.

Finkelstein, Norman. *Image and Reality of the Israel-Palestine Conflict*. New York: Verso, 1995.

Foer, Franklin. "Jeff Bezos's Master Plan." *Atlantic*, November 2019. https://www .theatlantic.com/magazine/archive/2019/11/what-jeff-bezos-wants/598363/.

Fogg, Martyn J. "Ethical Dimensions of Space Settlement." *Space Policy* 16, no. 3 (2000): 205–11. https://www.sciencedirect.com/science/article/pii/ S0265964600000242.

Fontenelle, Bernard le Bovier de. *Conversations on the Plurality of Worlds (1686)*. Translated by H. A. Hargreaves. Berkeley: University of California Press, 1990.

Foster, L. M., and Namrata Goswami. "What China's Antarctic Behavior Tells Us about the Future of Space." *Diplomat*, January 11, 2019.

Fox, Keolu, and Chanda Prescod-Weinstein. "The Fight for Mauna Kea Is a Fight against Colonial Science." *Nation*, July 24, 2019. https://www.thenation.com/ article/archive/mauna-kea-tmt-colonial-science/.

Francis. "Laudato Si': On Care for Our Common Home." Encyclical letter, May 24, 2015. http://www.vatican.va/content/francesco/en/encyclicals/documents/ papa-francesco_20150524_enciclica-laudato-si.html.

Friedman, Richard Elliott. *Who Wrote the Bible?* New York: Harper Collins, 1997.

Gorman, Alice. "The Cultural Landscape of Interplanetary Space." *Journal of Social Archaeology* 5, no. 1 (2005): 87–107.

Grinspoon, David. *Lonely Planets: The Natural Philosophy of Alien Life*. New York: Ecco, 2004.

Handmer, Annie. "MVA Public Forum on the Moon." Office of Other Spaces, Sat- ellite Applications Catapult UK, Moon Village Association, Space Junk Podcast, August 23, 2020. https://www.youtube.com/watch?v=8SB_ZwVgGOs.

Hanson, Robin. "How to Live in a Simulation." *Journal of Evolution and Technology* 7, no. 1 (September 2001). http://www.jetpress.org/volume7/simulation.htm.

Haroun, Fawaz, Shalom Ajibade, Philip Oladimeji, and John Kennedy Igbozurike. "Toward the Sustainability of Outer Space: Addressing the Issue of Space Debris." *New Space* 9, no. 1 (2021): 63–71.

Harvey, Graham, ed. *The Handbook of Contemporary Animism*. Durham, UK: Acumen, 2013.

Haskins, Caroline. "The Racist Language of Space Exploration." *Outline*, August 14, 2018. https://theoutline.com/post/5809/the-racist-language-of -space-exploration.

Healey, Robert M. "Jefferson on Judaism and the Jews: 'Divided We Stand, United, We Fall!'" *American Jewish History* 73, no. 4 (June 1984): 359–74.

Hore-Thorburn, Isabelle. "Trust Elon Musk to Make Going to Space Sound Shit." *Highsnobiety*, n.d.. https://www.highsnobiety.com/p/elon-musk-colonizing -mars-indentured-slavery/.

Howe, Daniel Walker. *What Hath God Wrought: The Transformation of America, 1815–1848*. Oxford: Oxford University Press, 2007.

Ingraham, Christopher. "A Proliferation of Space Junk Is Blocking Our View of the Cosmos, Research Shows." *Washington Post*, April 27, 2021.

Jacobs, Harrison. "Inside the Ultra-elite Explorers Club That Counts Jeff Bezos, Buzz Aldrin, and James Cameron as Members." *Business Insider*, December 30, 2017. https://www.businessinsider.com/explorers-club-new-york-elon-musk -james-cameron-buzz-aldrin-2017-12.

Jah, Morbia. "Acta non verba: That Should Be the Motto for NASA's Artemis Accords." *Aerospace America*, Jahniverse, July/August 2020. https://aerospace america.aiaa.org/departments/acta-non-verba-that-should-be-the-motto-for -nasas-artemis-accords/.

Jemisin, N. K. "Cloud Dragon Skies." In *How Long 'til Black Future Month?*, 113–25. New York: Orbit, 2018.

———. "The Ones Who Stay and Fight." In *How Long 'til Black Future Month?*, 1–13. New York: Orbit, 2018.

Kaçar, Betül. "Do We Send the Goo?" *Aeon*, November 21, 2020. https://aeon.co/ essays/if-were-alone-in-the-universe-should-we-do-anything-about-it.

Kaku, Michio. *The Future of Humanity: Terraforming Mars, Interstellar Travel, Immortality, and Our Destiny beyond Earth*. New York: Doubleday, 2018.

Kearns, Laurel, and Catherine Keller, eds. *Ecospirit: Religions and Philosophies for the Earth*. New York: Fordham University Press, 2007.

Kennedy, John F. "If the Soviets Control Space, They Can Control Earth." *Missiles and Rockets*, October 10, 1960, 12–13, 50.

———. "Special Message to the Congress on Urgent National Needs." May 25, 1961. John F. Kennedy Presidential Library and Museum. https://www.jfklibrary.org/ archives/other-resources/john-f-kennedy-speeches/united-states-congress -special-message-19610525.

Khatchadourian, Raffi. "The Elusive Peril of Space Junk." *New Yorker*, September 21, 2020. https://www.newyorker.com/magazine/2020/09/28/the-elusive -peril-of-space-junk.

Killian, James R., Jr. *Sputnik, Scientists, and Eisenhower: A Memoir of the First Special Assistant to the President for Science and Technology*. Cambridge, MA: MIT Press, 1977.

Kimmerer, Robin Wall. *Braiding Sweetgrass: Indigenous Wisdom, Scientific Knowledge, and the Teaching of Plants*. Minneapolis, MN: Milkweed, 2013.

Klinger, Julie Michelle. *Rare Earth Frontiers: From Terrestrial Subsoils to Lunar Landscapes*. Ithaca, NY: Cornell University Press, 2016.

Kminek, G., C. Conley, V. Hipkin, and H. Yano. Committee on Space Research. "COSPAR Planetary Protection Policy." December, 2017: https://cosparhq.cnes .fr/assets/uploads/2019/12/PPPolicyDecember-2017.pdf.

LaFleur, Ingrid, and Moriba Jah. "What Does the Afrofuture Say? W/Moriba Jah." Afrostrategy Strategies Institute, July 9, 2020. https://www.youtube.com/watch?v=B69ROBootPw.

Lampen, Claire. "Have TikTok Witches Actually 'Hexed the Moon'?" *New York*, The Cut, July 19, 2020. https://www.thecut.com/2020/07/some-tiktok-baby-witches-apparently-tried-to-hex-the-moon.html.

Langston, Scott M. "'A Running Thread of Ideals': Joshua and the Israelite Conquest in American History." In *On Prophets, Warriors, and Kings: Former Prophets through the Eyes of Their Interpreters*, edited by George J. Brooke and Ariel Feldman, 229–63. Berlin: De Gruyter, 2016.

Lazier, Benjamin. "Earthrise; or, the Globalization of the World Picture." *American Historical Review* 116, no. 3 (June 2011): 602–30.

Lee, John M. "'Silent Spring' Is Now Noisy Summer; Pesticides Industry up in Arms over a New Book. Rachel Carson Stirs Conflict—Producers Are Crying 'Foul.' Rachel Carson Upsets Industry." *New York Times*, July 22, 1962. https://www.nytimes.com/1962/07/22/archives/silent-spring-is-now-noisy-summer-pesticides-industry-up-in-arms.html.

Le Guin, Ursula K. "Newton's Sleep." In *A Fisherman of the Inland Sea: Stories*, 23–53. New York: Harper Perennial, 2005.

———. "The Ones Who Walk Away from Omelas." In *The Wind's Twelve Quarters*, 275–84. New York: William Morrow, 1975.

Lewis-Kraus, Gideon. "How the Pentagon Started Taking U.F.O.s Seriously." *New Yorker*, April 30, 2021. https://www.newyorker.com/magazine/2021/05/10/how-the-pentagon-started-taking-ufos-seriously.

Lovelock, James. *Gaia: A New Look at Life on Earth*. New York: Oxford University Press, 1979.

Margulis, Lynn, and Oona West. "Gaia and the Colonization of Mars." *GSA Today* 3, no. 11 (November, 1993): 277–80, 291.

Mather, Cotton. *Soldiers Counselled and Comforted: A Discourse Delivered unto Some Part of the Forces Engaged in the Just War of New-England against the Northern and Eastern Indians*. Boston: Samuel Green, 1689.

McKay, Christopher. "Planetary Ecosynthesis on Mars: Restoration Ecology and Environmental Ethics." In *Exploring the Origin, Extent, and Future of Life: Philosophical, Ethical, and Theological Perspectives*, edited by Constance M. Bertka, 245–60. Cambridge: Cambridge University Press, 2009.

Millard, Egan. "The Only Bible on the Moon Was Left There by an Episcopalian on Behalf of His Parish." Episcopal News Service, July 19, 2019. https://www.episcopalnewsservice.org/2019/07/19/the-only-bible-on-the-moon-was-brought-there-by-an-episcopalian-on-behalf-of-his-parish/.

Mohanta, Nibedita. "How Many Satellites Are Orbiting the Earth in 2021?" *Geospatial World*, May 28, 2021. https://www.geospatialworld.net/blogs/how-many-satellites-are-orbiting-the-earth-in-2021/.

Musk, Elon. "Making Humans a Multi-Planetary Species." *New Space* 5, no. 2 (2017): 46–61.

Nakahado, Sidney Nakao. "Should Space Be Part of a Development Strategy? Reflections Based on the Brazilian Experience." *New Space* 9, no. 1 (2021): 19–26.

National Aeronautics and Space Administration (NASA). "The Artemis Accords: Principles for Cooperation in the Civil Exploration and Use of the Moon, Mars,

Comets, and Asteroids for Peaceful Purposes." October 13, 2020. https://www
.nasa.gov/specials/artemis-accords/img/Artemis-Accords-signed-13Oct2020
.pdf.

———. "The Artemis Plan: Nasa's Lunar Exploration Program Overview." September 2020. https://www.nasa.gov/sites/default/files/atoms/files/artemis_plan
-20200921.pdf.

Newell, Catherine. "The Strange Case of Dr. Von Braun and Mr. Disney: Frontierland, Tomorrowland, and America's Final Frontier." *Journal of Religion and
Popular Culture* 25, no. 3 (Fall 2013): 416–29.

Nicholas V. "*Romanus Pontifex*: Granting the Portuguese a Perpetual Monopoly
in Trade with Africa." January 8, 1455. *Papal Encyclicals Online* https://www
.papalencyclicals.net/nichol05/romanus-pontifex.htm.

Nietzsche, Friedrich. *On the Genealogy of Morals.* Translated by Walter Kaufmann.
New York: Vintage Books, 1989.

O'Neill, Gerard K. *The High Frontier: Human Colonies in Space.* North Hollywood,
CA: Space Studies Institute Press, 1976.

"Orderly Formula." *Time,* October 28, 1957, 17–19.

Pace, Scott. "Space Development, Law, and Values (Lunch Keynote)." IISL Gal-
loway Space Law Symposium, December 13, 2017. https://spacepolicyonline
.com/wp-content/uploads/2017/12/Scott-Pace-to-Galloway-Symp-Dec-13
-2017.pdf.

Patterson, Richard Sharpe. *The Eagle and the Shield: A History of the Great Seal of
the United States.* Washington, DC: Office of the Historian, Bureau of Public
Affairs, Department of State, 1978. https://archive.org/details/TheEagleAnd
TheShield/page/n57/mode/2up.

Pence, Michael. "Address to the Fifth Meeting of the National Space Council."
March 26, 2019. https://www.youtube.com/watch?v=ZQkoFuNWXg8&t=
2027s.

Potter, Christopher. *The Earth Gazers: On Seeing Ourselves.* New York: Pegasus
Books, 2018.

Prior, Michael. *The Bible and Colonialism: A Moral Critique.* London: Bloomsbury,
1997.

———. "Confronting the Bible's Ethnic Cleansing in Palestine." *Link* 33, no. 5
(December 2000): 1–12.

———. "The Right to Expel: The Bible and Ethnic Cleansing." In *Palestinian Refugees: The Right of Return,* edited by Naseer Aruri, 9–35. London: Pluto, 2001.

Quinn, Adam G. "The New Age of Space Law: The Outer Space Treaty and the
Weaponization of Space." *Minnesota Journal of International Law* 17, no. 2
(2008): 475–502.

Ra, Sun. "We Travel the Spaceways." *Sun Ra and His Arkestra Greatest Hits.* https://
www.youtube.com/watch?v=oLn1JVsISho.

Rand, Ayn. *Atlas Shrugged.* New York: Dutton, 1992.

Rawls, Meredith L., Heidi B. Thiemann, Victor Chemin, Lucianne Walkowicz,
Mike W. Peel, and Yan G. Grange. "Satellite Constellation Internet Affordability and Need." *Research Notes of the AAS* 4, no. 189 (October 2020). https://
iopscience.iop.org/article/10.3847/2515-5172/abc48e.

Raz, Guy. "Lucianne Walkowicz: Should We Be Using Mars as a Backup Planet?"
TED Radio Hour, NPR, December 21, 2018. https://www.npr.org/transcripts/
678642121.

Resnick, Brian. "Apollo Astronauts Left Their Poop on the Moon." *Vox*, The Highlight, July 12, 2019. https://www.vox.com/science-and-health/2019/3/22/18236125/apollo-moon-poop-mars-science.

Robbins, Martin. "How Can Our Future Mars Colonies Be Free of Sexism and Racism?" *Guardian*, May 6, 2015. https://www.theguardian.com/science/the-lay-scientist/2015/may/06/how-can-our-future-mars-colonies-be-free-of-sexism-and-racism.

Roberge, Jack. "Elon Musk and Tesla: Saving the Planet by Being Awesome." *Villanovan*, February 4, 2020. http://www.villanovan.com/opinion/elon-musk-and-tesla-saving-the-planet-by-being-awesome/article_cf82b6d4-47bc-11ea-aa69-8b8a9ecb878a.html.

Rolston, Holmes, III. "The Preservation of Natural Value in the Solar System." In *Beyond Spaceship Earth: Environmental Ethics and the Solar System*, edited by Eugene C. Hargrove, 140–82. San Francisco: Sierra Club Books, 1987.

Rosenberg, Zach. "This Congressman Kept the U.S. and China from Exploring Space Together." *Complex*, December 17, 2013.

Roulette, Joey. "Elon Musk's Shot at Amazon Flares Monthslong Fight over Billionaires' Orbital Real Estate." *The Verge*, January 27, 2021. https://www.theverge.com/2021/1/27/22251127/elon-musk-bezos-amazon-billionaires-satellites-space.

Rubenstein, Mary-Jane. *Pantheologies: Gods, Worlds, Monsters*. New York: Columbia University Press, 2018.

———. *Worlds without End: The Many Lives of the Multiverse*. New York: Columbia University Press, 2014.

Russell, Andrew, and Lee Vinsel. "Whitey on Mars: Elon Musk and the Rise of Silicon Valley's Strange Trickle-Down Science." *Aeon*, February 1, 2017. https://aeon.co/essays/is-a-mission-to-mars-morally-defensible-given-todays-real-needs.

Sagan, Carl. *Cosmos*. New York: Ballantine Books, 1980.

———. *Pale Blue Dot: A Vision of the Human Future in Space*. New York: Random House, 1994.

Salaita, Steven. *The Holy Land in Transit: Colonialism and the Quest for Canaan*. Syracuse, NY: Syracuse University Press, 2006.

Schwartz, James S. J. "On the Moral Permissibility of Terraforming." *Ethics and the Environment* 18, no. 2 (Fall 2013): 1–31.

Seed, Patricia. *Ceremonies of Possession in Europe's Conquest of the New World, 1492–1640*. Cambridge: Cambridge University Press, 1995.

Sepulveda, Juan Ginés de. "Democrates Alter; or, on the Just Causes for War against the Indians" (1544). In *Boletín de la Real Academia de la Historia* 21 (October 1892). Originally translated for *Introduction to Contemporary Civilization in the West* (New York: Columbia University Press, 1946. http://www.columbia.edu/acis/ets/CCREAD/sepulved.htm.

Sheets, Michael. "FCC Approves SpaceX Change to Its Starlink Network, a Win Despite Objections from Amazon and Others." CNBC, Investing in Space, April 27, 2021. https://www.cnbc.com/2021/04/27/fcc-approves-spacex-starlink-modification-despite-objections.html.

Silko, Leslie Marmon. *Ceremony*. Anniversary ed. New York: Penguin, 2006.

Smolkin, Victoria. *A Sacred Space Is Never Empty: A History of Soviet Atheism*. Princeton, NJ: Princeton University Press, 2018.

Sparrow, Robert. "The Ethics of Terraforming." *Environmental Ethics* 21 (1999): 227–45.

Stamp, Jimmy. "American Myths:.Benjamin Franklin's Turkey and the Presidential Seal." *Smithsonian Magazine*, January 25, 2013. https://www.smithsonianmag .com/arts-culture/american-myths-benjamin-franklins-turkey-and-the -presidential-seal-6623414/.

Steinem, Gloria, and Sally Ride. "Sally Ride on the Future in Space." *Ms.*, January 1984, 86.

Stirone, Shannon. "Mars Is a Hellhole." *Atlantic*, February 26, 2021.

Stone, Peter. *1776*. 1972. Scripts.com. https://www.scripts.com/script-pdf/1574.

Szwed, John F. *Space Is the Place: The Lives and Times of Sun Ra*. New York: Da Capo, 1998.

Tavares, Frank, Denise Buckner, Dana Burton, Jordan McKaig, Parvathy Prem, Eleni Ravanis, Natalie Treviño, et al. "Ethical Exploration and the Role of Planetary Protection in Disrupting Colonial Practices." Cornell University, arXiv, October 15, 2020. https://arxiv.org/abs/2010.08344.

Thompson, Clive. "Monetizing the Final Frontier." *New Republic*, December 3, 2020. https://newrepublic.com/article/160303/monetizing-final-frontier.

Torres, Phil. "The Dangerous Ideas of Longtermism and Existential Risk." *Current Affairs*, July 28, 2021. https://www.currentaffairs.org/2021/07/the-dangerous -ideas-of-longtermism-and-existential-risk.

Treviño, Natalie B. "The Cosmos Is Not Finished." PhD diss., University of Western Ontario, 2020.

Trump, Donald J. "Remarks by President Trump in State of the Union Address." February 4, 2020. https://trumpwhitehouse.archives.gov/briefings-statements/ remarks-president-trump-state-union-address-3/.

Turner, Frederick Jackson. "The Significance of the Frontier in American History." In *The Frontier in American History*. New York: Henry Holt, 1935. Address delivered at the forty-first annual meeting of the State Historical Society of Wisconsin, December 14, 1893. http://xroads.virginia.edu/~HYPER/TURNER/.

Tutton, Richard. "Sociotechnical Imaginaries and Techno-Optimism: Examining Outer Space Utopias of Silicon Valley." *Science as Culture*, November 5, 2020, 1–24.

United Nations Office for Outer Space Affairs, Committee on the Peaceful Uses of Outer Space. "Agreement Governing the Activities of States on the Moon and Other Celestial Bodies (1984)." In *International Space Law: United Nations Instruments*, edited by United Nations Office for Outer Space Affairs, 30–39. New York: United Nations, 2017.

———. "Report of the Legal Subcommittee on Its Sixtieth Session, Held in Vienna from 31 May to 11 June 2021." August 2, 2021. https://www.unoosa.org/oosa/en/ oosadoc/data/documents/2021/aac.105/aac.1051243_0.html.

———. "Treaty on Principles Governing the Activities of States in the Exploration and Use of Outer Space, Including the Moon and Other Celestial Bodies (Outer Space Treaty) (1967)." In *International Space Law: United Nations Instruments*, edited by United Nations Office for Outer Space Affairs, 3–9. New York: United Nations, 2017.

United States Department of Defense. "Final Report on Organizational and Management Structure for the National Security Space Components of the Department of Defense." Department of Defense Report to Congressional

Defense Committees, August 9, 2018. https://media.defense.gov/2018/Aug/
09/2001952764/-1/-1/1/ORGANIZATIONAL-MANAGEMENT-STRUCTURE
-DOD-NATIONAL-SECURITY-SPACE-COMPONENTS.PDF.

United States Office of Space Commerce. "National Space Policy of the United
States of America." December 9, 2020. https://www.space.commerce.gov/
policy/national-space-policy/.

United States Space Force. *Spacepower: Doctrine for Space Forces.* Space Capstone
Publication, June 20, 2020. https://www.spaceforce.mil/Portals/1/Space
%20Capstone%20Publication_10%20Aug%202020.pdf.

Utrata, Alina. "Lost in Space." *Boston Review*, July 14, 2021. https://bostonreview
.net/science-nature/alina-utrata-lost-space.

Vance, Ashlee. *Elon Musk: Tesla, SpaceX, and the Quest for a Fantastic Future.* New
York: Ecco, 2017.

Vattel, Emmerich de. *The Law of Nations, or, Principles of the Law of Nature, Applied
to the Conduct and Affairs of Nations and Sovereigns (1758).* Indianapolis, IN:
Liberty Fund, 2008.

von Braun, Wernher. "For Space Buffs—National Space Institute—You Can Join."
Popular Science, May 1976, 72–73.

Wabuke, Hope. "Afrofuturism, Africanfuturism, and the Language of Black Speculative Literature." *Los Angeles Review of Books*, August 27, 2020. https://www
.lareviewofbooks.org/article/afrofuturism-africanfuturism-and-the-language
-of-black-speculative-literature/.

Walkowicz, Lucianne. "Let's Not Use Mars as a Backup Planet." TED, March
2015. https://www.ted.com/talks/lucianne_walkowicz_let_s_not_use_mars_as
_a_backup_planet?language=en.

Wall, Mike. "New Space Mining Legislation Is 'History in the Making.'" *Space.com*,
November 20, 2015, https://www.space.com/31177-space-mining-commercial
-spaceflight-congress.html.

Wark, K. "Wis2dom—Weaving Indigenous and Sustainability Sciences: Diversifying Our Methods Workshop." In *Weaving Indigenous and Sustainability Sciences:
Diversifying Our Methods*, edited by J. T. Johnson, R. P. Louis, and A Kliskey,
101–3. Arlington, VA: National Science Foundation, 2014.

Warrior, Robert Allen. "Canaanites, Cowboys, and Indians: Deliverance, Conquest, and Liberation Theology Today." *Christianity and Crisis*, September 11,
1989, 21–26.

Waxman, Olivia B. "Lots of People Have Theories about Neil Armstrong's 'One
Small Step for Man' Quote." *Time*, July 15, 2019. https://time.com/5621999/neil
-armstrong-quote/.

Welna, David. "Space Force Bible Blessing at National Cathedral Sparks Outrage."
NPR, National Security, January 13, 2020. https://www.npr.org/2020/01/13/
796028336/space-force-bible-blessing-at-national-cathedral-sparks-outrage.

"What Are We Waiting For?" *Collier's*, March 22, 1952.

Whitaker, Alexander. "Good Newes from Virginia (1613)." In *God's New Israel:
Religious Interpretations of American Destiny*, edited by Conrad Cherry, 30–36.
Chapel Hill: University of North Carolina Press, 1998.

White, Lynn, Jr.. "The Historical Roots of Our Ecologic Crisis." *Science* 155, no.
3767 (1967): 1203–7.

Wigglesworth, Michael. "God's Controversy with New England (1662)." In *God's

New Israel: Religious Interpretations of American Destiny, edited by Conrad Cherry, 42–53. Chapel Hill: University of North Carolina Press, 1998.

Wilkins, John. *The Discovery of a World in the Moone; or, a Discourse to Prove That 'Tis Probable There May Be Another Habitable World in That Planet*. London: Michael Sparke and Edward Forrest, 1638.

Wynter, Sylvia. "The Pope Must Have Been Drunk, the King of Castile a Madman: Culture as Actuality, and the Caribbean Rethinking Modernity." In *Reordering of Culture: Latin America, the Caribbean and Canada in the Hood*, edited by Alvina Ruprecht and Cecilia Taiana, 17–41. Ottawa, ON: Carleton University Press, 1995.

Young, M. Jane. "'Pity the Indians of Outer Space': Native American Views of the Space Program." *Western Folklore* 46, no. 4 (October 1987): 269–79.

Youngquist, Paul. *A Pure Solar World: Sun Ra and the Birth of Afrofuturism*. Austin: University of Texas Press, 2016.

Zimmerer, Jürgen. "The Birth of the Ostland out of the Spirit of Colonialism: A Postcolonial Perspective on the Nazi Policy of Conquest and Extermination." *Patterns of Prejudice* 39, no. 2 (2005): 197–219.

Zubrin, Robert. "Why We Humans Should Colonize Mars!" *Theology and Science* 17, no. 3 (2019): 305–16.

———. "Wokeists Assault Space Exploration." *National Review*, November 14, 2020. https://www.nationalreview.com/2020/11/wokeists-assault-space -exploration/.

Zubrin, Robert, and Richard Wagner. *The Case for Mars: The Plan to Settle the Red Planet and Why We Must*. New York: Free Press, 2011.

Index

Page numbers followed by "*f*" refer to figures.